T0185485

QUANTUM THEORY AS EMERGENT PHENOMENON

The Statistical Mechanics of Matrix Models as the Precursor of Quantum Field Theory

Quantum mechanics is our most successful physical theory. However, it raises conceptual issues that have perplexed physicists and philosophers of science for decades. This book develops a new approach, based on the proposal that quantum theory is not a complete, final theory, but is in fact an emergent phenomenon arising from a deeper level of dynamics. The dynamics at this deeper level is taken to be an extension of classical dynamics to non-commuting matrix variables, with cyclic permutation inside a trace used as the basic calculational tool. With plausible assumptions, quantum theory is shown to emerge as the statistical thermodynamics of this underlying theory, with the canonical commutation–anticommutation relations derived from a generalized equipartition theorem. Brownian motion corrections to this thermodynamics are argued to lead to state vector reduction and to the probabilistic interpretation of quantum theory, making contact with recent phenomenological proposals for stochastic modifications to Schrödinger dynamics.

STEPHEN L. ADLER received his Ph.D. degree in theoretical physics from Princeton. He has been a Professor in the School of Natural Sciences at the Institute for Advanced Study since 1969, and from 1979 to 2003 held the State of New Jersey Albert Einstein Professorship there.

Dr. Adler's research has included seminal papers in current algebras, sum rules, perturbation theory anomalies, and high energy neutrino processes. Dr. Adler has also done important work on neutral current phenomenology, strong field electromagnetic processes, acceleration methods for Monte Carlo algorithms, induced gravity, non-Abelian monopoles, and models for quark confinement. For nearly twenty years he has been studying embeddings of standard quantum mechanics in larger mathematical frameworks, with results described in this volume.

Dr. Adler is a member of the National Academy of Sciences, and is a Fellow of the American Physical Society, the American Academy of Arts and Sciences, and the American Association for the Advancement of Science. He received the J. J. Sakurai Prize in particle phenomenology, awarded by the American Physical Society, in 1988, and the Dirac Prize and Medal awarded by the International Center for Theoretical Physics in Trieste, in 1998.

QUANTUM THEORY AS AN EMERGENT PHENOMENON

PHENOMENON

The Statistical Mechanics of Matrix Models
as the Precursor of Quantum Field Theory

STEPHEN L. ADLER

Institute for Advanced Study, Princeton

CAMBRIDGE UNIVERSITY PRESS
Cambridge, New York, Melbourne, Madrid, Cape Town, Singapore, São Paulo, Delhi

Cambridge University Press
The Edinburgh Building, Cambridge CB2 8RU, UK

Published in the United States of America by Cambridge University Press, New York

www.cambridge.org
Information on this title: www.cambridge.org/9780521115971

First published 2004
This digitally printed version 2009

A catalogue record for this publication is available from the British Library

Library of Congress Cataloguing in Publication data
Adler, Stephen L.
Quantum theory as an emergent phenomenon: the statistical dynamics of global unitary
invariant matrix models as the precursor of quantum field theory / Stephen L. Adler.
p.   cm.
Includes bibliographical references and index.
ISBN 0 521 83194 6
1. Quantum theory.   I. Title: Statistical dynamics of global unitary invariant matrix models
as the precursor of quantum field theory.   II. Title.
QC174.12.A32 2004
530.12–dc22   2003064019

ISBN 978-0-521-83194-9 hardback
ISBN 978-0-521-11597-1 paperback

To Sarah Brett-Smith, with love and admiration

Contents

Acknowledgements

I have many people to thank for their assistance in aspects of this work. The discovery by my thesis student Andrew Millard of the conservation of $\tilde{C}$ provided the underpinning for the entire project. I am greatly indebted to him, and to my other collaborators in the course of parts of this work, Gyan Bhanot, Dorje Brody, Todd Brun, Larry Horwitz, Lane Hughston, Achim Kempf, Indrajit Mitra, John Weckel, and Yong-Shi Wu. I am grateful to Jeeva Anandan, Angelo Bassi, Todd Brun, Lajos Diósi, Larry Horwitz, Lane Hughston, Gerald Goldin, Stanley Liu, Peter Morgan, Philip Pearle, Artem Starodubtsev, and several anonymous publisher's reviewers, for many insightful comments on the first draft of this book. I am particularly indebted to Philip Pearle for detailed comments on the Introduction and Chapter 6, to Larry Horwitz for a careful reading of the entire manuscript, and to Todd Brun and Peter Morgan for remarks that led to the present form of Sec. 5.6. Finally, I wish to thank my wife Sarah for her perceptive support throughout this long project.

I have benefited from conversations and/or email correspondence with a great many others as well; a list (undoubtedly incomplete) includes: Philip Anderson, John Bahcall, Vijay Balasubramanian, Lowell Brown, Jeremy Butterfield, Tian-Yu Cao, Sudip Chakravarty, Freeman Dyson, Sheldon Goldstein, GianCarlo Ghirardi, Siyuan Han, William Happer, James Hartle, Roman Jackiw, Abraham Klein, John Klauder, Pawan Kumar, Joel Lebowitz, Anthony Leggett, James Lukens, G. Mangano, Herbert Neuberger, Ian Percival, Michael Ramalis, Soo-Jong Rey, Lee Smolin, Yuri Suhov, Leo Stodolsky, Terry Tao, Charles Thorn, Sam Treiman, Walter Troost, Steven Weinberg, Frank Wilczek, David Wineland, and Edward Witten.

Parts of this book are based on papers that were previously published in *Nuclear Physics B* (Adler, 1994; Adler and Millard, 1996; Adler, 1997a) and in *Physics Letters B* (Adler, 1997b; Adler and Horwitz, 2003), and I wish to thank Elsevier Science, Ltd. for permission to use this material. I similarly wish to thank Institute of Physics Publishing Ltd. for permission to use material originally published in

Journal of Physics A: Math. Gen. (Adler, 2002). Finally, I wish to acknowledge the American Physical Society for use of material originally published in papers appearing in the *Journal of Mathematical Physics* (Adler, Bhanot, and Weckel, 1994; Adler, 1998; Adler and Kempf, 1998; Adler and Horwitz, 1996, 2000) and in *Physical Review D* (Adler and Wu, 1994; Adler, 2000, 2003a).

I also wish to acknowledge the hospitality of the Aspen Center for Physics, and of both the Department of Applied Mathematics and Theoretical Physics and Clare Hall at Cambridge University, as well as my home base at the Institute for Advanced Study in Princeton. The Albert Einstein Professorship that I held while writing this book was partially funded by the State of New Jersey, and my work is also supported in part by the Department of Energy under Grant No. DE-FG02-90ER40542.

Introduction and overview

Quantum mechanics is our most successful physical theory. It underlies our very detailed understanding of atomic physics, chemistry, and nuclear physics, and the many technologies to which physical systems in these regimes give rise. Additionally, relativistic quantum mechanics is the basis for the standard model of elementary particles, which very successfully gives a partial unification of the forces operating at the atomic, nuclear, and subnuclear levels.

However, from its inception the probabilistic nature of quantum mechanics, and the fact that "quantum measurements" in the orthodox formulation appear to require the intervention of non-quantum mechanical "classical systems," have led to speculations by many physicists, mathematicians, and philosophers of science that quantum mechanics may be incomplete. Among the Founding Fathers of quantum theory, Einstein and Schrödinger were both of the opinion that quantum mechanics is in some way unsatisfactory, and this view has been amplified in more recent profound work of John Bell, among others. In an opposing camp, many others in the physics, mathematics, and philosophy communities have attempted to provide an interpretational foundation in which quantum mechanics remains a complete and self-contained system. Among the Founding Fathers, Bohr, Born, and Heisenberg maintained that quantum mechanics is a complete system, and a number of recent proposals have been made to improve upon or to provide alternatives to their "Copenhagen Interpretation." The debate continues, and has spawned an enormous literature. While it is beyond the scope of this book to give a detailed review of all the proposals that have been made, to set the stage we give a brief discussion of the measurement problem in Section 1, and we survey some of the current proposals to revise the interpretational foundation of quantum mechanics in Section 2.

The rest of this book, however, is based on the premise that quantum mechanics is in fact not a complete system, but rather represents a very accurate asymptotic approximation to a deeper level of dynamics. Motivations for pursuing this track are given in Section 3. The detailed proposal to be developed in this book is that *quantum mechanics is not a complete theory, but rather is an emergent phenomenon arising from the statistical mechanics of matrix models that have a global unitary invariance.* We use "emergent" here in the sense that it is used in condensed matter, molecular dynamics, and complex systems theory, where higher level phenomena (phonons, superconductivity, fluid mechanics, etc.) are seen to arise or "emerge" as the expressions, in appropriate dynamical contexts, of an underlying dynamics that at first glance shows little resemblance to these phenomena. Initial ideas in this direction were developed by the author and collaborators in a number of papers dealing with the properties of what we termed "generalized quantum dynamics" or, in the terminology that we shall use in this exposition,"trace dynamics." The purpose of this book is to give a comprehensive review of this earlier work, with a number of significant additions and modifications that bring the project closer to its goal. We shall also relate our proposal to a substantial body of literature on stochastic modifications of the Schrödinger equation, which we believe provides the low energy phenomenology, expressed in terms of experimentally accessible observables, for the pre-quantum dynamics that we develop here. A quick overview of what we intend to accomplish in the subsequent chapters is given in Section 4, and some brief remarks on the history of this project are given in Section 5.

Certain sections of this book are more technical in that they involve some knowledge of supersymmetry techniques and, although included for completeness, are not essential to follow the main line of development; these are marked with an asterisk (*) in the section head. The exposition of the text is based on dynamical variables that are matrices in complex Hilbert space, but many of the ideas carry over to a statistical dynamics of matrix models in real or quaternionic Hilbert space, as sketched in Appendix A. Discussions of other topics needed to keep our treatment self-contained are given in further appendices, and our notational conventions are reviewed in the introductory paragraphs preceding Appendix A.

1 The quantum measurement problem

Quantum mechanics works perfectly well in describing microscopic phenomena, and even in describing phenomena in which many particles act coherently in one or a small number of quantum states, as in Bose–Einstein condensates, superfluids, and superconducting Josephson junctions. Conceptual problems arise only when one tries to apply the rules of quantum mechanics simultaneously to a microscopic system and to the macroscopic apparatus that is measuring the state of the microscopic system; this is the origin of the notorious "quantum measurement problem." We shall give here a simplified, "bare bones" description of the measurement problem, taking as an example a variant of the familiar Stern–Gerlach experiment. (For a selection of papers on the measurement problem, see the reprint volume Wheeler and Zurek, 1983.)

Consider a source emitting spin-1/2 particles with polarized spins, so that all particles have spin component up along the x axis; that is, the initial beam is in a state with $S_x = 1/2$. (We shall see in a moment how this is accomplished in practice.) The particles then go through an inhomogeneous magnetic field aligned along the z axis, which splits the beam into two spatially displaced components, corresponding to components of the beam with spin component $S_z = 1/2$ and $S_z = -1/2$, as shown in Fig. 1a. The quantum mechanical description of what has happened so far is simply the spin state decomposition (with appropriate phase conventions)

$$|S_x = 1/2\rangle = \frac{1}{\sqrt{2}}(|S_z = 1/2\rangle + |S_z = -1/2\rangle). \tag{1a}$$

At this point *no measurement has been made*; if we pass the split beams through a second inhomogeneous field with the direction of inhomogeneity reversed, as in Fig. 1b, and devote great care to the isolation of the beams from environmental influences, the two components of the beam merge back into one and what emerges from the combined apparatus is the original state $|S_x = 1/2\rangle$. (An analysis of issues involved in achieving spin coherence, and further references, are given in Sculley, Englert, and Schwinger, 1989.)

To make a measurement, one must intercept one or both beams with a macroscopic measuring apparatus that absorbs the beam and registers a count in some form. When the measuring apparatus A intercepts both beams, we get the conventional Stern–Gerlach setup pictured in Fig. 1c. This is described, in the von Neumann (1932) model of measurement, by the evolution of the initial state $|S_x = 1/2\rangle|A_{\text{initial}}\rangle$ into a state in which the measured system and the apparatus are entangled

$$\frac{1}{\sqrt{2}}(|S_z = 1/2\rangle|A_+\rangle + |S_z = -1/2\rangle|A_-\rangle), \tag{1b}$$

where $|A_+\rangle$ is an apparatus state with a count shown on the upper counter and none on the lower counter, while $|A_-\rangle$ is an apparatus state with a count shown on the lower counter and none on the upper counter.

Once an apparatus intervenes in this way, two salient features become apparent. The first is that it is impossible in practice to coherently recombine the total system consisting of beam and apparatus so as to regain the initial state $|S_x = 1/2\rangle$. This feature, that the two legs of the apparatus have decohered, can be understood within the framework of quantum mechanics: since the apparatus state is a complex, large system, reversing the joint evolution of beams and apparatus with sufficient accuracy to preserve interference requires an unachievable control over the apparatus state. This is all the more so because in general the apparatus is in interaction with an external environment, into which phase coherence information is rapidly dissipated, making a coherent recombination of the beams a practical impossibility. In density matrix language, the off-diagonal components of the density matrix, when traced over the internal states of the apparatus and the environment, rapidly vanish because of decoherence effects, leaving just diagonal components that represent the probabilities for seeing the apparatus register an up or a down S_z spin component. (For further discussions of decoherence theory, see Harris and Stodolsky, 1981; Joos and Zeh, 1985; Zurek, 1991; and Joos, 1999.)

The second salient feature is that while there are definite probabilities for the apparatus to register a spin up or a spin down component, the outcome of any given run of a particle through the apparatus cannot be predicted; part of the time it registers in the "up" counter, and part of the time it registers in the "down" counter. (In the above example, the probabilities for registering "up" and "down" are both $1/2$, but for general orientations of the apparatus axis the probabilities will be $\sin^2 \theta/2$ and $\cos^2 \theta/2$, with θ the angle by which the inhomogeneous magnetic field is rotated with respect to the x axis.) This unpredictability of individual outcomes is the origin of the quantum measurement problem. If we maintain that quantum mechanics should apply to both the particle passing through the apparatus *and* to the measuring apparatus itself, then the final state at time t is described by a unitary evolution $U = \exp(-iHt)$ applied to the initial state, and this describes a superposition as in Eq. (1b), not an either–or choice between outcomes that are described by orthogonal states in Hilbert space. Since environmental decoherence effects still involve a unitary evolution (in an enlarged Hilbert space describing the system, apparatus, and environment), they cannot account for this either–or choice observed in the experimental outcomes. (See Adler, 2003b for a more detailed discussion of this point, and for extensive literature references. For an opposing viewpoint, see the review of Zurek, 2003.)

It is not necessary for the apparatus to intercept both beams for a measurement problem to be apparent. Consider the apparatus illustrated in Fig. 1d, which intercepts only the "down" leg of the experiment. If the particles are gated into the apparatus at definite time intervals, then a count on the "down" meter indicates that a particle has been detected there, and subsequent downstream measurements in the "up" leg will detect no particle there. If there is no count on the "down" meter (i.e., a "down" meter anti-coincidence), then one can say with certainty that the particle has passed through the "up" leg of the apparatus and is in a polarized state $|S_z = 1/2\rangle$; this is how one produces a polarized beam. Decoherence accounts for

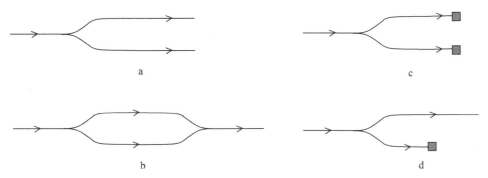

Figure 1 Beam paths through variants of the Stern-Gerlach experiment. Where the beams separate or recombine, there are magnetic fields that are not shown. a. Spin up and down components are separated and continue to propagate. b. Spin up and down components are separated, propagate, and then are coherently recombined. c. Spin up and down components are separated and each impinges on a detector. d. Spin up and down components are separated, the down component impinges on a detector, while the up component continues to propagate, producing a spin up polarized beam.

the fact that we cannot in practice reconstitute the original state $|S_x = 1/2\rangle$, but it cannot account for the stochastic pattern in which polarized particles emerge from the "up" leg of our apparatus.

There are two conventional ways to try to avoid the measurement dilemma just stated. The first is to assert that quantum mechanics has only a statistical interpretation, and should only be applied to describe the statistical properties of multiple repetitions of an experiment, but not to any individual run. However, with the advent of our ability to trap individual particles for long periods, and to manipulate their quantum states (e.g., the particle emerging from the "up" beam in Fig. 1d could be run into a trap, and manipulated there), this interpretation of quantum mechanics becomes dubious. The second is to adopt the Copenhagen interpretation, and to state by fiat that the unitary state vector evolution of quantum mechanics does not apply to measurement situations. One then adds to the unitary evolution postulate a second postulate, that of state vector reduction, which states that after a measurement one sees a unit normalized state corresponding to the measurement outcome $|f\rangle$, with a probability given by the Born rule $P_f = |\langle f|\Psi\rangle|^2$ as applied to the initial state $|\Psi\rangle$ being measured.

While perfectly consistent for all experiments that have been performed to date, the Copenhagen interpretation is at odds with the our belief that quantum mechanics should have universal applicability, and should describe the behavior of large systems (such as a measuring apparatus) as well as microscopic ones. It also has the bizarre feature of erecting a probabilistic theory, without an underlying sample space of individual events, the coarse-grained behavior of which is described by the probabilities. In all other applications of probability theory, probabilities emerge from the fact that one cannot observe, or chooses not to observe, individual details which deterministically specify the outcomes. Quantum mechanics is unique in that probabilities (or in some formulations, expectation values) are introduced as a postulate, without emerging by some well-defined rule from an underlying sample space of predictable individual events.

There are two logical possibilities for dealing with the problems just sketched. The first is to maintain that quantum mechanics is exactly correct, but in need of an improved conceptual foundation. One way to do this is to generalize the Copenhagen interpretation, so as to eliminate the apparently arbitrary distinction between "system" and "apparatus," and to give a set of extended interpretive rules with general applicability. This is the goal of the "consistent histories" approach to quantum foundations. Another way to do this is to extend the kinematic rules of quantum mechanics so as to give a concrete specification of a hidden sample space, that is constructed so as to be in principle unobservable, which leads to Born rule probabilities because full details of the sample space cannot be seen. This is what is done in certain versions of the "many worlds" approach, and in the Bohmian and Ax–Kochen approaches to quantum theory.

The second logical alternative is to consider the possibility that quantum mechanics is only a very accurate approximation to a deeper level of dynamics, which in turn gives a unified understanding of both unitary Schrödinger evolution and measurement dynamics. In this case the sample space that is created is not constructed so as to be unobservable, and detectable deviations from quantum mechanics become possible, leading to experimental constraints on the model parameters. As in any approach that proceeds by creating a sample space, there are so-called "hidden variables," and so important constraints imposed by no-go theorems coming from the work of Kochen and Specker (1967), Bell (1964, 1987), and others, have to be observed.

In Section 2 immediately following, we shall briefly describe the approaches that proceed from the assumption that quantum theory is exact but requires a new conceptual foundation. In Section 3 we shall give motivations for considering the possibility that quantum mechanics is in fact not an exact, final theory, which leads into the main themes of this book.

2 Reinterpretations of quantum mechanical foundations

A number of approaches to the reinterpretation of quantum foundations, assuming that quantum theory is exact, have been explored in recent years. Our aim in this section is to give a brief overview with entry points to the relevant literature, without attempting either a detailed exposition or a critique.

2.1 Histories

The histories approach is a generalization of the Copenhagen interpretation, that replaces the imprecise notions of an "apparatus" and a "measurement" with more precise concepts based on histories. The basic objects in this approach are time-dependent projectors $E_k(t_k)$ associated with events (defined as properties at given times) occurring in a history, and the probability of a history is then postulated to be given by

$$p = \text{Tr}[E_n(t_n) \ldots E_1(t_1)\rho E_1(t_1) \ldots E_n(t_n)], \qquad (2a)$$

with ρ the initial density matrix. This definition, supplemented by the notion of a family of decohering histories, which describes mutually exclusive evolutions with probabilities that sum to unity, can be argued to lead to all of the usual properties of quantum mechanical probabilities. In this interpretation, state vector reduction appears only as a Bayesian statistical rule for relating the density matrix after a measurement to that before the measurement. Detailed accounts of the histories approach can be found in the book of Griffiths (2002), the review and books of Omnès (1992, 1994, 1999), and the lectures of Hartle (1992). The histories approach involves no enlargement of the basic mathematical apparatus of quantum mechanics, and may still be relevant as a detailed description of quantum behavior even if quantum mechanics turns out to be an approximation to a deeper level of dynamics.

The three approaches that we discuss next all enlarge the mathematical structure of quantum mechanics, so as to create a sample space which forms the basis for the probabilistic interpretation. However, in all three cases the attributes that distinguish "individuals" in the sample space are not observable, so that there are no predictions that differ from those of standard quantum mechanics. Because these theories reproduce the results of quantum mechanics, it is evident that the assumptions of the Kochen and Specker (1967) and Bell (1964) no-go results are evaded. In the Bell case, for example, this results from nonlocality in the construction of the hidden sample space.

2.2 Bohmian mechanics

In Bohmian mechanics (Bohm, 1952), in addition to the Schrödinger equation for the N-body wave-function $\psi(q_1, \ldots, q_N, t)$ that obeys

$$i\hbar \frac{\partial \psi}{\partial t} = \left(-\sum_{k=1}^{N} \frac{\hbar^2}{2m_k} \nabla_{q_k}^2 + V \right) \psi, \qquad (2b)$$

one enlarges the mathematical framework by introducing hidden "particles" moving in configuration space with coordinates Q_k and velocities

$$v_k = \frac{dQ_k}{dt} = \frac{\hbar}{m_k} \text{Im} \nabla_{Q_k} \log \psi(Q_1, \ldots, Q_N, t). \tag{2c}$$

The state of the individual system is then specified by giving both the wave function and the coordinates Q_k of the hidden particles. If the probability in configuration space is assumed to obey the Born rule $p = |\psi|^2$ at some initial time, the Bohmian equations then imply that this continues to be true at all subsequent times. Arguments have been given that the Bohmian initial time probability postulate follows from considerations of "typicality" of initial configurations. For detailed expositions, see Bub (1997), Dürr, Goldstein, and Zanghi (1992), and Dürr, Goldstein, Tumulka, and Zanghi (2003).

2.3 The Ax–Kochen proposal

Ax and Kochen (1999) extend the mathematical framework of quantum theory to encompass the "individual," by identifying the ray with the quantum ensemble, and the ray representative, i.e., the $U(1)$ phase associated with a particular state vector, with the individual. They then give a mathematical construction to specify a unique physical state from knowledge of the toroid of phases. They argue that if the a priori distribution of phases is assumed to be uniform, then their construction implies that the probabilities of outcomes obey the usual Born rule.

2.4 Everett's "many worlds" interpretation

In the "many worlds" interpretation introduced by Everett (1957), there is no state vector reduction, but only Schrödinger evolution of the entire universe. In this interpretation, to describe N successive quantum measurements requires consideration of an N-fold tensor product wave function. The mathematical framework can be enlarged to create a sample space by considering the space of all possible such tensor products, and defining a suitable measure on this space. This procedure, given in the De Witt and Graham (1973) versions of many worlds, is the basis for arguments obtaining the Born rule as the probability for the occurrence of a particular outcome, that is, as the probability of finding oneself on a particular branch of the universal wave function.

Since the reinterpretations of quantum theory sketched here all aim, by construction, to reproduce the entire body of predictions of nonrelativistic quantum theory, they cannot be experimentally falsified (unless deviations from quantum theory are eventually established). Thus, apart from issues of the extent to which they can be generalized to encompass relativistic quantum field theory, the choice between them is somewhat a matter of taste. Rather than join in the already extensive literature debating their strengths and weaknesses, we shall proceed now to consider an alternative possibility, that quantum mechanics is in fact not an exact, complete structure.

3 Motivations for believing that quantum mechanics is incomplete

As surveyed in the preceding section, one approach to the quantum measurement problem and associated "paradoxes" of quantum theory is to continue to assume that quantum mechanics is exactly correct, and to attempt to supply it with a new foundational interpretation. However, there is another logical possibility, which is to suppose that quantum mechanics is not exactly correct, but represents an extremely accurate approximation to a qualitatively different level of dynamics. Since quantum theory is an extraordinarily successful physical theory, one can ask why try to replace it with something else? We respond to this question by listing a number of motivations for considering the possibility that quantum mechanics, and quantum field theory, may require modification at a deeper level.

3.1 Historical precedent

The historical development of physics contains many examples of theories that seemed to be exact in the context for which they were developed, only to require modification when applied to a larger arena of phenomena. Newtonian mechanics and Galilean relativity appeared to be exact in the context of planetary orbits, until the need for their special and general relativistic extensions became apparent in the early twentieth century. Classical predictability appeared to be exact in the context of classical mechanics, thermodynamics, and statistical mechanics, until confronted with the problems of the blackbody radiation spectrum and the discreteness of spectral lines at the end of the nineteenth century. The Landau mean field theory of critical phenomena was considered to be exact, until confronted with experimental data showing anomalous critical scaling, requiring the modern Kadanoff–Fisher–Wilson theory of critical phenomena for its explanation. Given these historical precedents, there seems to be no compelling reason to assume that quantum mechanics is immune to the general rule, that theories are only valid within a given regime, and may require modification when extended beyond that regime.

3.2 The quantum measurement problem

As we have discussed in Section 1, the unitary evolution of standard quantum mechanics does not describe what happens when measurements are made, but conventionally has to be supplemented by an additional postulate of nonunitary state vector reduction when a "measurement" is performed by a "classical" apparatus. As many authors have stressed, an economical resolution of the measurement "paradoxes" would be achieved if one could find a more fundamental underlying dynamics, from which the unitary evolution and the state vector reduction aspects of conventional quantum mechanics would emerge in a natural way in the appropriate physical contexts. Such a resolution should show in a natural way why quantum mechanics is probabilistic, by endowing it with an underlying sample space, and should show how probabilities become actualities for individual outcomes.

3.3 What is the origin of "canonical quantization"?

The standard approach to constructing a quantum field theory consists in first writing down the corresponding classical theory, and then "quantizing" it by reinterpreting the classical quantities as operators, and replacing the classical Poisson brackets by $-i/\hbar$ times the corresponding commutators or anticommutators. However, since quantum theory is more fundamental than classical theory, it seems odd that one has to construct it by starting from the classical limit; the canonical quantization approach has very much the flavor of an algorithm for inverting the classical limit of quantum mechanics. Moreover, it is known through the theorem of Groenewold and van Hove (for a recent review, see Giulini, 2003) that the Dirac recipe of replacing Poisson brackets by commutators cannot consistently be applied to general polynomials in the canonical variables, but only to the restricted class of second-order polynomials. Additionally, what is the origin of Planck's constant $\hbar$? One might hope that in a new theory underlying quantum mechanics, one would work with operators from the outset and proceed directly to operator equations of motion without first starting from the classical limit, and that one would also achieve an understanding of why there is a fundamental quantum of action.

3.4 Infinities and nonlocality

An outstanding problem in quantum mechanics (or more specifically, in quantum field theory) is the presence of infinities arising from the local structure of the canonical commutation/anticommutation relations, and an outstanding puzzle in quantum mechanics is the nonlocality seen, for example, in Einstein, Podolsky, and Rosen (1935) type experiments. Both of these considerations motivate many studies that have been made of quantum foundations, and in our view suggest that quantum mechanics may arise from a deeper level of physics that is substantially nonlocal.

3.5 Unification of quantum theory with gravitation

There are a number of indications that conventional quantum field theory must be modified in a profound fashion in order for it to be successfully combined with gravitational physics. In generic curved spacetimes, it is not possible to give a precise formulation of the particle production rate, nor is there necessarily a well-defined concept of conserved energy. As is well known, when conventionally quantized, general relativity leads to a non-renormalizable quantum field theory. Another indication that quantum field theory must be modified when combined with gravitational physics is provided by recent ideas on "holography," which suggest that the association of degrees of freedom with volume subdivisions must break down near the Planck energy. These problems are among the motivations for replacing quantum field theory by a quantized theory of strings, but it is possible that modification of the rules of quantum theory will also be needed to give a fully successful unification of the forces. In other words, in addition to exploring "pre-geometrical" theories to explain quantum gravity, one may have to explore "pre-quantum mechanical" theories as well.

3.6 The cosmological constant

Another indication that quantum mechanics may have to be modified to deal with gravitational phenomena is provided by the problem of the cosmological constant. In conventional quantum field theory it is very hard to understand why the observed cosmological constant is 120 orders of magnitude smaller than the natural scale provided by the Planck energy. Either unbroken scale invariance or unbroken supersymmetry would forbid the appearance of a cosmological constant, but they also forbid the appearance of a realistic particle mass spectrum, and so in conventional quantum theory they do not provide a basis for solving the cosmological constant problem. The difficulty that arises here can be formulated as a mismatch between the single constraint needed – a sum rule dictating the vanishing of the cosmological constant – and the infinite number of constraints arising from having conserved operator scale and conformal transformation generators or a conserved operator supercurrent. One possible way to resolve the cosmological constant problem would be to find a deeper level of theory, in which the single constraint needed to resolve the cosmological constant problem is matched, in a naive counting sense, to the constraint arising from imposing scale invariance or supersymmetry on that deeper level.

3.7 A concrete proposal

Last, but not least, we have a concrete proposal for *how* to replace quantum mechanics by a deeper level of physical theory, that will have significant implications for all of the issues just listed. Our proposal, already noted in the introductory paragraphs, is that quantum mechanics is an emergent phenomenon arising from the statistical mechanics of matrix models with a global unitary invariance. To be more specific now, our idea is to start from a classical dynamics in which the dynamical variables are non-commutative matrices or operators. (We will use the terms matrix and operator interchangeably throughout this book, and do not commit ourselves as to whether they are finite $N \times N$ dimensional, or infinite dimensional as obtained in the limit $N \to \infty$.) Despite the non-commutativity, a sensible Lagrangian and Hamiltonian dynamics is obtained by forming the Lagrangian and Hamiltonian as traces of polynomials in the dynamical variables, and repeatedly using cyclic permutation under the trace, which restricts the dynamical variables to be "trace class," and is the motivation for calling the resulting dynamics "trace dynamics." We further assume that the Lagrangian and Hamiltonian are constructed without use of non-dynamical matrix coefficients, so that there is an invariance under simultaneous, identical unitary transformations of all of the dynamical variables, that is, there is a global unitary invariance. We assume that the complicated dynamical equations resulting from this system rapidly reach statistical equilibrium, and then show that with suitable approximations, the statistical thermodynamics of the canonical ensemble for this system takes the form of quantum field theory. Specifically, the statistical thermodynamics of the underlying trace dynamics leads to the usual canonical commutation/anticommutation algebra of quantum mechanics, as well as the Heisenberg time evolution of operators, and these in turn, imply the usual rules of Schrödinger picture quantum mechanics. The requirements for the underlying trace dynamics to yield quantum theory at the level of thermodynamics are stringent, and include both the generation of a mass hierarchy and the existence of boson–fermion balance. We cannot at this point give *the* specific theory that obeys all of the needed conditions; this is a topic for future work. There may of course be no theory that satisfies our conditions, but our hope is that there will be at least one underlying theory that fits into the general framework developed here.

The proposal just sketched corresponds to the relations between classical mechanics, quantum mechanics, and the underlying "trace dynamics" theory that is qualitatively pictured in Fig. 2. At the top level is classical mechanics, for which the dynamical variables are all commutative. Classical dynamical variables are usually represented as ordinary numbers, but they can also be represented as matrices in a Hilbert space, in which case they must all be taken as proportional to the unit matrix. Through the canonical quantization procedure one arrives at the middle level of quantum mechanics and quantum field theory, from which one recovers classical mechanics by taking a classical limit in which (passing over many subtleties) Planck's constant effectively approaches zero. In quantum mechanics the dynamical canonical coordinate and momentum variables are a special class of infinite matrices which obey the canonical commutation/anticommutation relations. Our proposal is that there is another level, more basic than quantum mechanics, governed by a global unitary invariant trace dynamics. Here the dynamical variables are completely general matrices, with no a priori assumption of commutativity properties. From the equilibrium statistical mechanics of trace dynamics, the rules of quantum mechanics emerge as an approximate thermodynamic description of the behavior of low energy phenomena. "Low energy" here means small relative to the natural energy scale implicit in the canonical ensemble for trace dynamics, which we identify with the Planck scale, and by "equilibrium" we mean local equilibrium, permitting spatial variations associated with dynamics at the low energy scale. Brownian motion corrections to the thermodynamics of trace dynamics then lead to fluctuation corrections to quantum mechanics which take the form of stochastic modifications of the Schrödinger equation, that can account in a mathematically precise way for state vector reduction with Born rule probabilities.

The remainder of this book consists of a detailed development of the ideas just outlined and diagrammed in Fig. 2.

4 An overview of this book

As a guide to the reader, we give here a brief overview of the book.

In Chapter 1 we introduce our notation for the non-commutative matrices that form the dynamical variables of trace dynamics. Bosonic variables are represented by ordinary complex matrices, while fermionic variables are represented by complex Grassmann matrices. We then give the basic bilinear and trilinear cyclic trace identities that are used in subsequent derivations. We next show, by using the cyclic invariance of the trace of a polynomial (or more generally, a meromorphic function) in the dynamical variables, that one can consistently define an operator which gives the derivative of a trace quantity with respect to an operator. Using this operator derivative, we formulate a trace dynamics analog of classical Lagrangian and Hamiltonian dynamics, which gives a *classical* dynamics of matrix models, and we show that in this dynamics the trace Hamiltonian $\mathbf{H} = \mathrm{Tr}H$ is conserved. We construct a generalized Poisson bracket appropriate to trace dynamics, discuss its properties, and give some applications. Finally, we contrast the dynamical equations for the non-commuting matrices of trace dynamics with the unitary

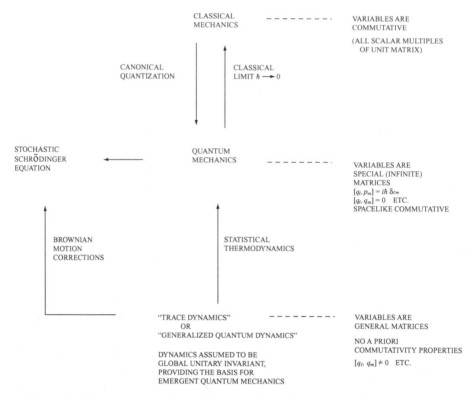

Figure 2 Diagrammatic relations between the various theories discussed in this book: classical mechanics, quantum mechanics and quantum field theory, trace dynamics (also called generalized quantum dynamics), and stochastically modified Schrödinger picture quantum mechanics.

evolution obtained by assuming a Heisenberg picture dynamics, in which the dynamical variables obey the usual canonical commutators/anticommutators of quantum mechanics.

In Chapter 2 we explore further conserved quantities in trace dynamics. We show that when there are equal numbers of fermionic canonical coordinate and momentum factors in each term in the trace Hamiltonian, then there is a conserved trace fermion number **N**. We next consider the class of trace dynamics models that are global unitary invariant, that is, have a trace Hamiltonian that is constructed from the matrix dynamical variables using only c-number coefficients, thus excluding the use of non-dynamical matrices as coefficients. For this class of models, we show that there is a conserved operator with the dimensions of action, which we call $\tilde{C}$, which is equal to the sum of bosonic commutators $[q, p]$ minus the corresponding sum of fermionic anticommutators $\{q, p\}$, and which is the conserved matrix-valued Noether charge corresponding to the assumed global unitary invariance. This operator plays a fundamental role in our argument for an emergent quantum mechanics. With the usual fermionic adjointness assignment $\tilde{C}$ is anti-self-adjoint (but for alternative adjointness assignments $\tilde{C}$ can have a self-adjoint part, which we assume, if present, to be very small). We proceed to give the four-current analogs of $\mathbf{N}$ and $\tilde{C}$ when the trace Lagrangian is specialized to describe continuum spacetime theories, and also discuss the trace energy-momentum tensor $T_{\mu\nu}$, which is a conserved quantity when the underlying trace dynamics is Poincaré invariant. In this case the conserved charge $\tilde{C}$ is also Poincaré invariant, which explains why later on, when we assume that the low energy statistical thermodynamics is dominated by the $\tilde{C}$ term in the canonical ensemble, with the $\mathbf{H}$ term, which defines the preferred frame implicit in the canonical ensemble, effectively decoupled, a Poincaré invariant quantum field structure emerges. As a simple illustrative example of the trace dynamics formalism we consider the model in which a Dirac fermion matrix field is coupled to a scalar Klein–Gordon matrix field. Finally, we discuss the symmetry properties of the conserved quantities under interchange of fermionic canonical coordinates and momenta.

In Chapter 3 (which can be omitted on a first reading), we continue the discussion of specific models that illustrate the formalism of trace dynamics, this time in the context of theories with global supersymmetry. In succession, we discuss the trace dynamics analogs of the Wess–Zumino model, the supersymmetric Yang–Mills model, and the so-called "matrix model for M-theory." Finally, we briefly describe difficulties encountered in attempting to extend this discussion to theories, such as supergravity, with local supersymmetry.

In Chapter 4 we begin the analysis of the statistical mechanics of matrix models. We open by pointing out how our procedure differs from conventional approaches to matrix models (see, e.g., Brézin and Wadia, 1993), in which the classical dynamics of these models is canonically quantized. By contrast, in developing an emergent quantum theory we treat the classical dynamics of matrix models as fundamental, and analyze its consequences by using an appropriate generalization of statistical mechanics. To introduce statistical methods, we first define a natural measure for matrix phase space, and show that this measure obeys a generalized Liouville theorem. This then allow us to apply statistical mechanical methods, in which we maximize the entropy subject to constraints, to derive the canonical ensemble for trace dynamics, in which the generic conserved quantities $\mathbf{H}$, $\mathbf{N}$, and $\tilde{C}$ appear multiplied by Lagrange multipliers that represent generalized "temperatures." At this point we specialize the ensemble to one that has maximal symmetry consistent with the ensemble average $\langle \tilde{C} \rangle_{\text{AV}}$ being non-zero, which we show implies that $\langle \tilde{C} \rangle_{\text{AV}}$ can be written as $i_{\text{eff}} \hbar$, with i_{eff} an anti-self-adjoint matrix with square -1, and with $\hbar$ the real positive factor defined by this polar decomposition of $\langle \tilde{C} \rangle_{\text{AV}}$. The matrix i_{eff} will play the role of i in our argument for an emergent quantum theory and, as suggested by the notation, $\hbar$ will play the role of the reduced Planck's constant. We continue the statistical analysis by showing that the canonical ensemble can also be derived by starting from the microcanonical ensemble, and considering the equilibrium of a large subsystem in contact with a much larger "bath." We then give a discussion (which can be omitted on a first reading) of gauge fixing in the canonical ensemble for trace dynamics models with a local gauge invariance. Finally, we discuss the implications of the fact that the canonical ensemble only partially breaks the assumed global unitary invariance; this analysis plays an important role in establishing, in the next chapter, a correspondence between canonical ensemble averages in trace dynamics and Wightman functions in an emergent quantum field theory. First we formulate the need for a global unitary fixing in general terms, and then (in a section which can be omitted on first reading) give a detailed construction of global unitary fixings for the partition function.

Chapter 5 contains the heart of our argument for the emergence of quantum field theory from trace dynamics. The basic observation, developed through the detailed derivations of this chapter, is that since the conserved operator $\tilde{C}$ is a sum of bosonic commutators minus a sum of fermionic anticommutators, the *equipartitioning* of $\tilde{C}$ in canonical ensemble averages leads to an effective canonical commutator (anticommutator) structure for the bosonic (fermionic) dynamical operator variables. We proceed in analogy with the standard equipartition theorems of statistical mechanics, which we show can be viewed as simple Ward identities. We begin by deriving a general Ward identity for trace dynamics, and showing that its structure can be augmented by varying external source terms in the canonical ensemble. We then show that if we make a low energy approximation, in which we assume that the underlying trace Hamiltonian (or Lagrangian) is such that there is a decoupling of contributions arising from variation of the **H** term in the canonical ensemble, so that the averaged dynamics is dominated by the $\tilde{C}$ term, the structure of quantum theory emerges. The reason that dynamical information can be extracted from equilibrium averages is that the trace dynamics equations of motion, in Hamiltonian form, take the first-order form $\dot{x} = F_x$, with x a particular phase space variable and with F_x an operator function of all of the phase space variables. Hence by showing that, within our approximations, the canonical ensemble average of F_x times a universal constant, in the presence of sources, is equal to the corresponding canonical ensemble average of $[x, H]$, we learn that $\dot{x}$ is equivalently given by the usual Heisenberg evolution formula of quantum mechanics. The universal constant, which plays the role of i times the reduced Planck constant in the emergent quantum theory, is given by the ensemble average $\langle \tilde{C} \rangle_{\mathrm{AV}}$. In Chapter 4, this quantity was represented in polar form as $i_{\mathrm{eff}} \hbar$, with i_{eff} a matrix square root of -1 and with the parts of the dynamical variables that commute with i_{eff} identified as the effective canonical variables of the emergent quantum theory. With these identifications, a correspondence between canonical ensemble averages in trace dynamics, and Wightman functions in an emergent quantum field theory, can be established. We note that, although polynomials in the dynamical variables in general depend of the choice of unitary fixing imposed in Chapter 4, the Wightman functions and more generally transition probabilities can be expressed in terms of trace quantities that are independent of the unitary fixing. An examination of alternative Ward identities shows that our decoupling approximation involves nontrivial constraints on the behavior of the underlying theory, including certain support properties in operator phase space, and a requirement of boson–fermion balance which strongly hints at a need for supersymmetry. Up to this point the emergent quantum theory is in the Heisenberg picture; we then proceed to derive the Schrödinger equation for the emergent quantum theory. Finally, we discuss the Kochen–Specker (1967) and Bell (1964) "no-go" arguments for hidden variable theories, and show how their assumptions are evaded by our statistical mechanical argument for an emergent quantum theory.

In Chapter 6 we analyze Brownian motion corrections to the emergent quantum theory, thereby making contact with a long line of investigations of phenomenological stochastic Schrödinger equations pioneered by Pearle (1976, 1979, 1984, 1989), Ghirardi, Rimini, and Weber (1986), Ghirardi, Pearle, and Rimini (1990), Gisin (1984, 1989), Diósi (1988a,b, 1989), and Percival (1994). Making simple models for the form of the fluctuation terms in the Ward identities arising from $\tilde{C}$, we give scenarios for deriving the standard localization and energy-driven stochastic Schrödinger equations. We then review the proof that these equations are members of a general class of stochastic equations that leads to state vector reduction with Born rule probabilities, and review the formulas needed to estimate reduction rates in the energy-driven and localization models. We discuss the phenomenology of the energy-driven equation, giving constraints on its stochastic parameter coming from current experiments, and giving a critical survey of mechanisms that have been proposed to produce the energy dispersion needed for rapid state vector reduction in measurement contexts. We finally briefly survey the phenomenology of the localization approach, referring the reader to the recent reviews of Bassi and Ghirardi (2003) and Pearle (1999b) for a more detailed treatment. We conclude that as of this writing the localization model is favored, both because the assumptions needed to derive it within our framework are more robust, and because there are unresolved problems with the mechanisms that have been proposed to explain reduction in the energy-driven model.

Finally, in Chapter 7 we indicate how our proposal for an emergent quantum theory addresses the motivational questions raised above in Section 3, and discuss some of the issues that will be relevant for future developments. We again emphasize here that, while we have given a general framework in which an emergent quantum theory may appear, we have not identified the specific theory in which all our requirements are realized.

We conclude this overview by noting work of other authors that also considers the premise that quantum mechanics may be modified at a deeper level. Both 't Hooft (1988, 1997, 1999a,b, 2001a,b, 2002, 2003) and Smolin (1983, 1985, 2002) have proposed models for the emergence of quantum theory from an underlying level of dynamics. While their basic philosophy is very similar to that of this book, the details of what they do differs substantially, and neither the statistical mechanical canonical ensemble nor the conserved operator $\tilde{C}$ play a role in their analyses. 't Hooft proposes that beneath quantum theory there is a deterministic classical, chaotic dynamics, with a set of attractors that determine the effective emergent quantum theory. Smolin considers classical matrix models, with an explicit stochastic noise along the lines of that used by Nelson (1969, 1985) giving rise to the quantum behavior. Despite the evident differences, there may be elements of their approaches that will ultimately be seen to share common ground with ours. At a phenomenological level, Bialynicki-Birula and Mycielski (1976) and Weinberg (1989a,b,c) have considered nonlinear, deterministic modifications of the Schrödinger equation, and comparison of their models with experiment (Bollinger *et al.*, 1991) sets strong bounds on such possible modifications to conventional quantum theory. Their models have been shown by Polchinski (1991), Gisin (1989, 1990), and Gisin and Rigo (1995), to have the problem of predicting superluminal signal propagation. When we discuss phenomenological modifications of the Schrödinger equation in Chapter 6, the only nonlinearities will appear in fluctuating, stochastic terms, for which the experimental bounds are very weak, and which do not give rise to superluminal signal propagation.

5 Brief historical remarks on trace dynamics

I close this introductory chapter with some brief historical remarks on trace dynamics, and on how the proposal that it can serve as a foundation for quantum theory came about.

First of all, the idea of using a trace variational principle to generate operator equations goes back to the inception of quantum mechanics; see Born and Jordan (1925), who in Section 2 of their paper introduce a symbolic differentiation of operator monomials under a trace that is identical to the bosonic case of the one used here. They did not develop this idea further, and it remained unnoticed for many years. A Hamiltonian variational principle based on this idea was later used by Kerman and Klein (1963) to generate equations of motion for many-body physics. I am indebted to A. Klein for bringing these references to my attention several years ago; see Klein, Li, and Vassanji (1980) and Greenberg *et al.* (1996) for further references to many-body theory applications. None of these early references erect the full apparatus for trace dynamics constructed here in Chapters 1 and 2.

The idea of using the operator derivative of a trace as the basis for formulating a new dynamical theory, as opposed to as a tool for studying standard quantum theory, first appeared in a paper (Adler, 1979) in which I made an unsuccessful attempt to formulate a dynamics for the Harari–Shupe preon model. I subsequently returned to trace dynamics, under the name "generalized quantum dynamics," in conjunction with the writing of my book *Quaternionic Quantum Mechanics and Quantum Fields* (Adler, 1995) for two reasons. First of all, I was unable to find any extension of the canonical quantization procedure to quaternionic Hilbert space, and so was led to study trace dynamics as a way to generate operator equations of motion directly, without canonically quantizing a classical theory. In this connection, the Hamiltonian version of trace dynamics and the generalized Poisson bracket were formulated (Adler, 1994), and the Jacobi identity for the generalized Poisson bracket was subsequently proved by Adler, Bhanot, and Weckel (1994). Secondly, an anonymous publisher's reviewer for the 1995 book raised the issue of whether quaternionic Hilbert space might ameliorate the measurement problems of quantum mechanics. The answer turned out to be "no," because quaternionic quantum theory simply substitutes quaternion unitary for complex unitary Schrödinger evolution, and so the need for a separate state vector reduction postulate persists. Investigating this issue suggested, however, that trace dynamics, which is not equivalent to a unitary evolution, might lead to a resolution of the measurement problem. However, further development of this notion required a way to get back from the more general trace dynamics to quantum mechanics. The attempts to do this in Section 13.6 of the 1995 book only worked for one degree of freedom, and did not have an obvious extension to systems with many degrees of freedom, although in hindsight the discussion of Eqs. (13.90a–f) of that section anticipated the form of the conserved operator $\tilde{C}$. (For a recent paper along similar lines, see Starodubtsev, 2002.)

At this point a crucial ingredient was supplied by Millard (personal communication, 1995), who as part of a thesis investigation (Millard, 1997) of trace dynamics theories with Weyl-ordered Hamiltonians, discovered the existence of the conserved operator $\tilde{C}$. Its structure was immediately suggestive of an equipartition argument for the emergence of quantum mechanics from trace dynamics, and this was developed in detail in the paper of Adler and Millard (1996), which provides the basis for much of the material in Chapters 2 through 5 of this book. Further progress was made in papers with other collaborators, in particular Adler and Horwitz (1996), which constructed the microcanonical ensemble for trace dynamics and used this to rederive the canonical ensemble, and Adler and Kempf

(1998), which reexpressed the general argument for conservation of $\tilde{C}$ given by Adler and Millard (1996) in terms of global unitary invariance, gave a group-theoretic characterization of the maximally symmetric canonical ensemble, and showed that there is a consistency requirement of boson–fermion balance. A key remaining obstacle was that in the paper of Adler and Millard (1996), the canonical ensemble averages of products of dynamical variables associated with spatial points were identified with vacuum expectations of operators in the emergent field theory, and with this putative correspondence it was not possible to establish the Wightman spectral condition. In the spring of 2001, I revisited the entire program, and discovered the need to take account of the fact that the canonical ensemble does not fully break the assumed global unitary invariance, as noted in Adler and Kempf (1998) and as discussed in detail in Section 4.5. Thus the unrestricted canonical ensemble averages correspond to traces, rather than vacuum expectations, of operator products, which is why there was an obstacle to identifying them with Wightman functions. When the integrations defining the canonical ensemble are restricted to break this residual unitary invariance, it becomes possible to set up a consistent correspondence between the trace dynamics canonical ensemble averages of operator products, and the vacuum expectations of the corresponding operator products in the emergent quantum theory; my discovery of this fact, as well as other technical progress made in the course of the 2001 research, led to the decision to write this book. The full details of the global unitary fixing, given in Section 4.6 and Appendix G, were worked out in Adler and Horwitz (2003) during the final stages of my work on the book manuscript.

Finally, I make an historical and notational comment on the method by which fermions are introduced into the theory. In all of the papers in the trace dynamics program before 1997, fermions were introduced through a $(-1)^F$ operator insertion in the trace, rather than by use of a Grassmann algebra as done in Adler and Kempf (1998) and in this book. The principal results of the older work are unaffected by this change, but certain details are altered. Also, in this book we consistently use an adjoint convention in which two Grassmann odd grade matrices χ_1 and χ_2 obey $(\chi_1\chi_2)^\dagger = -\chi_2^\dagger\chi_1^\dagger$. This convention is implicit in Adler and Horwitz (2003), but the older papers, such as Adler (1997a,b), as well as the first draft of this book that appeared on the Los Alamos archive as hep-th/0206120, use a convention in which $(\chi_1\chi_2)^\dagger = \chi_2^\dagger\chi_1^\dagger$. The results of this book (except for Appendix G) can be readily expressed in this second convention by the inclusion of additional factors of i in various places.

1

Trace dynamics: the classical Lagrangian and Hamiltonian dynamics of matrix models

In this chapter we set up a classical Lagrangian and Hamiltonian dynamics for matrix models. The fundamental idea is to set up an analog of classical dynamics in which the phase space variables are non-commutative, and the basic tool that allows one to accomplish this is cyclic invariance under a trace. Since no assumptions about commutativity of the phase space variables (such as canonical commutators/anticommutators) are made at this stage, the dynamics that we set up is not the same as standard quantum mechanics. Quantum mechanical behavior will be seen to emerge only when, in Chapters 4 and 5, we study the statistical mechanics of the classical matrix dynamics formulated here.

In Section 1.1, we introduce our basic notation for bosonic and fermionic matrices, and give the cyclic identities that will be used repeatedly throughout the book. In Section 1.2, we define the derivative of a trace quantity with respect to an operator, and give the basic properties of this definition. In Section 1.3, we use the operator derivative to formulate a Lagrangian and Hamiltonian dynamics for matrix models. In Section 1.4, we introduce a generalized Poisson bracket appropriate to trace dynamics, constructed from the operator derivative defined in Section 1.2, and give its properties and some applications. Finally, in Section 1.5 we discuss the relation between the trace dynamics time evolution equations, and the usual unitary Heisenberg picture equations of motion obtained when one assumes standard canonical commutators/anticommutators.

1.1 Bosonic and fermionic matrices and the cyclic trace identities

We shall assume finite-dimensional matrices, although ultimately an extension to the infinite-dimensional case may be needed. The matrix elements of these matrices will be constructed from ordinary complex numbers, and from complex anticommuting Grassmann numbers. Just as a complex number can be decomposed into real and imaginary parts, $c = c_R + i c_I$ with $c_{R,I}$ real, a complex Grassmann number can be decomposed into real and imaginary parts, $\chi = \chi_R + i\chi_I$ with $\chi_{R,I}$ real. Real Grassmann numbers are built up as products of a basis of real Grassmann elements $\chi_1, \chi_2, \ldots$ which obey the anticommutative algebra $\{\chi_r, \chi_s\} = 0$, an algebra which implies in particular that the square of any Grassmann element vanishes (for a further discussion of Grassmann algebras and references, see the introduction to the Appendices). Clearly a product of an even number of Grassmann elements commutes with all elements of the Grassmann algebra, while the product of an odd number of Grassmann elements anticommutes with any other product constructed from an odd number of Grassmann element factors. Thus the Grassmann algebra divides into two sectors: unity, together with the all products of an even number of Grassmann elements, form what is called the even grade sector of the Grassmann algebra, while all products of an odd number of Grassmann elements form what is called the odd grade sector of the algebra. Any even grade element commutes with any even or odd grade element, while two odd grade elements anticommute with one another. Grassmann elements are a familiar feature in the field theory literature on path integrals and supersymmetry, where even grade Grassmann elements represent bosonic fields, while odd grade Grassmann elements represent fermionic fields. Even or odd grade Grassmann elements can be combined with complex number coefficients; we will then speak of even or odd grade elements of the Grassmann algebra over the complex numbers.

Let B_1 and B_2 be two $N \times N$ matrices with matrix elements that are even grade elements of a Grassmann algebra over the complex numbers, and let Tr be the ordinary matrix trace, which obeys the cyclic property

$$\text{Tr}B_1 B_2 = \sum_{m,n} (B_1)_{mn}(B_2)_{nm} = \sum_{m,n} (B_2)_{nm}(B_1)_{mn} = \text{Tr}B_2 B_1. \qquad (1.1a)$$

Similarly, let χ_1 and χ_2 be two $N \times N$ matrices with matrix elements that are odd grade elements of a Grassmann algebra over the complex numbers, which anticommute rather than commute, so that the cyclic property for these takes the form

$$\text{Tr}\chi_1 \chi_2 = \sum_{m,n} (\chi_1)_{mn}(\chi_2)_{nm} = -\sum_{m,n} (\chi_2)_{nm}(\chi_1)_{mn} = -\text{Tr}\chi_2 \chi_1. \qquad (1.1b)$$

Since the even and odd grade elements of a Grassmann algebra over the complex numbers commute, one has a final bilinear cyclic identity

$$\text{Tr}B\chi = \text{Tr}\chi B. \qquad (1.1c)$$

We shall refer to the Grassmann even and Grassmann odd matrices B, χ as being of bosonic and fermionic type, respectively. Clearly, operators that are of mixed bosonic and fermionic type can always be linearly decomposed into components that are purely bosonic or purely fermionic in character.

The extra minus sign that appears in the odd grade case of Eq. (1.1b) has implications for the adjoint properties of matrices. Letting $\mathcal{O}^g$ be a matrix of grade g, we define the adjoint by

$$(\mathcal{O}^{g\,\dagger})_{mn} = (\mathcal{O}^g)^*_{nm}, \tag{1.1d}$$

that is, irrespective of the grade g, we define the matrix adjoint as the complex conjugate $*$ of the matrix with row and column labels transposed. Letting now $\mathcal{O}^{g_1}_1$ and $\mathcal{O}^{g_2}_2$ be two matrices of grade g_1 and g_2 respectively, this definition implies that

$$\begin{aligned}
(\mathcal{O}^{g_1}_1 \mathcal{O}^{g_2}_2)^\dagger_{mn} &= (\mathcal{O}^{g_1}_1 \mathcal{O}^{g_2}_2)^*_{nm} = \sum_k (\mathcal{O}^{g_1}_1)^*_{nk}(\mathcal{O}^{g_2}_2)^*_{km} \\
&= (-1)^{g_1 g_2} \sum_k (\mathcal{O}^{g_2}_2)^*_{km}(\mathcal{O}^{g_1}_1)^*_{nk} = (-1)^{g_1 g_2} \sum_k (\mathcal{O}^{g_2\,\dagger}_2)_{mk}(\mathcal{O}^{g_1\,\dagger}_1)_{kn} \\
&= (-1)^{g_1 g_2}(\mathcal{O}^{g_2\,\dagger}_2 \mathcal{O}^{g_1\,\dagger}_1)_{mn} \tag{1.1e}
\end{aligned}$$

so that as a matrix statement we have

$$(\mathcal{O}^{g_1}_1 \mathcal{O}^{g_2}_2)^\dagger = (-1)^{g_1 g_2} \mathcal{O}^{g_2\,\dagger}_2 \mathcal{O}^{g_1\,\dagger}_1. \tag{1.1f}$$

Thus, two odd grade matrices χ_1 and χ_2 obey the adjoint rule $(\chi_1 \chi_2)^\dagger = -\chi_2^\dagger \chi_1^\dagger$, and a general string of matrices obeys the rule

$$(\mathcal{O}^{g_1}_1 \ldots \mathcal{O}^{g_n}_n)^\dagger = (-1)^{\sum_{i<j} g_i g_j} \mathcal{O}^{g_n\,\dagger}_n \ldots \mathcal{O}^{g_1\,\dagger}_1. \tag{1.1g}$$

The difference between the adjoint convention used in this book, and one in which there is no grading factor in Eqs. (1.1e–g), is discussed in the introduction to the Appendices.

The cyclic/anticyclic properties of Eqs. (1.1a–c) are the basic identities from which further cyclic properties can be derived. For example, from the basic bilinear identities one immediately derives the trilinear cyclic identities

$$\begin{aligned}
&\text{Tr}\,B_1[B_2, B_3] = \text{Tr}\,B_2[B_3, B_1] = \text{Tr}\,B_3[B_1, B_2], \\
&\text{Tr}\,B_1\{B_2, B_3\} = \text{Tr}\,B_2\{B_3, B_1\} = \text{Tr}\,B_3\{B_1, B_2\}, \\
&\text{Tr}\,B\{\chi_1, \chi_2\} = \text{Tr}\,\chi_1[\chi_2, B] = \text{Tr}\,\chi_2[\chi_1, B], \\
&\text{Tr}\,\chi_1\{B, \chi_2\} = \text{Tr}\,\{\chi_1, B\}\chi_2 = \text{Tr}\,[\chi_1, \chi_2]B, \\
&\text{Tr}\,\chi[B_1, B_2] = \text{Tr}\,B_2[\chi, B_1] = \text{Tr}\,B_1[B_2, \chi], \\
&\text{Tr}\,\chi\{B_1, B_2\} = \text{Tr}\,B_2\{\chi, B_1\} = \text{Tr}\,B_1\{B_2, \chi\}, \\
&\text{Tr}\,\chi_1\{\chi_2, \chi_3\} = \text{Tr}\,\chi_2\{\chi_3, \chi_1\} = \text{Tr}\,\chi_3\{\chi_1, \chi_2\}, \\
&\text{Tr}\,\chi_1[\chi_2, \chi_3] = \text{Tr}\,\chi_2[\chi_3, \chi_1] = \text{Tr}\,\chi_3[\chi_1, \chi_2],
\end{aligned} \tag{1.2}$$

which are used repeatedly in trace dynamics calculations. In these equations, and throughout the text, $[X, Y] \equiv XY - YX$ denotes a matrix commutator, and $\{X, Y\} = XY + YX$ a matrix anticommutator.

1.2 Derivative of a trace with respect to an operator

The basic observation of trace dynamics (Born and Jordan, 1925; Adler, 1994, 1995) is that, given the trace of a polynomial P constructed from non-commuting matrix or operator variables (we shall use the terms "matrix" and "operator" interchangeably throughout the book), one can *define* a derivative of the complex number $\mathrm{Tr}\,P$ with respect to an operator variable $\mathcal{O}$ by varying and then cyclically permuting so that in each term the factor $\delta\mathcal{O}$ stands on the right. This gives the fundamental definition

$$\delta\mathrm{TrP} = \mathrm{Tr}\frac{\delta\mathrm{Tr}P}{\delta\mathcal{O}}\delta\mathcal{O}, \tag{1.3a}$$

or in the condensed notation that we shall use henceforth, in which $\mathbf{P} \equiv \mathrm{Tr}P$

$$\delta\mathbf{P} = \mathrm{Tr}\frac{\delta\mathbf{P}}{\delta\mathcal{O}}\delta\mathcal{O}, \tag{1.3b}$$

which for arbitrary infinitesimal $\delta\mathcal{O}$ defines the operator $\delta\mathbf{P}/\delta\mathcal{O}$. In general we will take $\mathcal{O}$ to be either of bosonic or fermionic (but not of mixed) type, and we will construct $\mathbf{P}$ to always be an even grade element of the Grassmann algebra. (When P is fermionic, we can always make it bosonic by multiplying it by a c-number auxiliary Grassmann element α.) With these restrictions, for $\delta\mathcal{O}$ of the same type as $\mathcal{O}$, the operator derivative $\delta\mathbf{P}/\delta\mathcal{O}$ will be of the same type as $\mathcal{O}$, that is, either both will be bosonic or both will be fermionic. Although we have introduced Eqs. (1.3a,b) for polynomials P, the definition immediately extends to functions expressible as power series in polynomials, and by use of the operator identity $\delta X^{-1} = -X^{-1}\delta X X^{-1}$, to meromorphic functions of polynomials in the dynamical variables as well.

Let us illustrate the fundamental definition of Eqs. (1.3a,b) with some simple examples. Suppose that P is a bosonic monomial containing only a single factor of the operator $\mathcal{O}$, so that P has the form

$$P = A\mathcal{O}B, \tag{1.3c}$$

with A and B operators that in general do not commute with each other or with $\mathcal{O}$. Then when $\mathcal{O}$ is varied, the corresponding variation of P is $\delta P = A(\delta\mathcal{O})B$, and so cyclically permuting B to the left we have

$$\begin{aligned}\delta\mathrm{Tr}P &= \epsilon_B\mathrm{Tr}BA\delta\mathcal{O}, \\ \frac{\delta\mathbf{P}}{\delta\mathcal{O}} &= \epsilon_B BA,\end{aligned} \tag{1.3d}$$

where $\epsilon_B = 1$ when the operator B is bosonic, and where $\epsilon_B = -1$ when the operator B is fermionic. Note that since we are taking P to be bosonic, the operator product $A\mathcal{O}$ is of the same bosonic or fermionic type as B, so we have $\epsilon_B = \epsilon_{A\mathcal{O}}$ and could equally well write

$$\frac{\delta \mathbf{P}}{\delta \mathcal{O}} = \epsilon_{A\mathcal{O}} BA, \tag{1.3e}$$

which is the result that we would obtain by cyclically permuting $A\delta\mathcal{O}$ to the right in the expression for $\delta \mathrm{Tr} P$. As a second illustration, suppose that P is a bosonic monomial containing two factors of the operator $\mathcal{O}$ that is being varied, and so has the general structure

$$P = A\mathcal{O}B\mathcal{O}C, \tag{1.3f}$$

with A, B, and C operators that in general do not commute with each other or with $\mathcal{O}$. Then applying the chain rule of differentiation, when $\mathcal{O}$ is varied the corresponding variation of P is $\delta P = A(\delta\mathcal{O})B\mathcal{O}C + A\mathcal{O}B(\delta\mathcal{O})C$. Thus we have in this case

$$\delta \mathrm{Tr} P = \mathrm{Tr}\big(\epsilon_{A\mathcal{O}} B\mathcal{O}CA(\delta\mathcal{O}) + \epsilon_C C A\mathcal{O}B(\delta\mathcal{O})\big),$$
$$\frac{\delta \mathbf{P}}{\delta \mathcal{O}} = \epsilon_{A\mathcal{O}} B\mathcal{O}CA + \epsilon_C C A\mathcal{O}B, \tag{1.3g}$$

with $\epsilon_C = 1(-1)$ according as whether C is bosonic (fermionic), and with $\epsilon_{A\mathcal{O}} = 1(-1)$ according as whether the product $A\mathcal{O}$ is bosonic (fermionic). The generalization to the case when P contains $N_\mathcal{O}$ factors of $\mathcal{O}$ follows the same pattern, with δP now consisting of a sum of $N_\mathcal{O}$ terms, in each of which a different factor $\mathcal{O}$ is varied. In each of these terms, the factors are then cyclically permuted so that $\delta\mathcal{O}$ stands on the right, identifying $\big($by comparison with Eq. (1.3b)$\big)$ the contribution of the term in question to $\delta\mathbf{P}/\delta\mathcal{O}$.

The definition of Eq. (1.3b) has the important property that if δP vanishes for arbitrary variations $\delta\mathcal{O}$ of the same type as $\mathcal{O}$, then the operator derivative $\delta\mathbf{P}/\delta\mathcal{O}$ must vanish. To see this, let us expand $\delta\mathbf{P}/\delta\mathcal{O}$ in the form

$$\frac{\delta \mathbf{P}}{\delta \mathcal{O}} = \sum_n C_n K_n, \tag{1.4a}$$

with the K_n distinct Grassmann monomials that are all c-numbers (i.e., multiples of the $N \times N$ unit matrix), and with the C_n complex matrix coefficients that are unit elements in the Grassmann algebra. Let us choose $\delta\mathcal{O}$ to be an infinitesimal α times $C_p^\dagger$, with α a real number when $\mathcal{O}$ is bosonic, and with α a Grassmann element not appearing in K_p when $\mathcal{O}$ is fermionic. $\big($There must be at least one such element, or else K_p would make an identically vanishing contribution to Eq. (1.3b), and could not appear in the sum in Eq. (1.4a).) We then have

$$0 = \sum_n \mathrm{Tr} C_p^\dagger C_n K_n \alpha, \tag{1.4b}$$

and since the coefficients of all distinct Grassmann monomials must vanish separately, we have in particular

$$0 = \mathrm{Tr} C_p^\dagger C_p. \tag{1.4c}$$

This implies the vanishing of the matrix coefficient C_p, and letting p range over all index values appearing in the sum in Eq. (1.4a), we conclude that

$$\frac{\delta \mathbf{P}}{\delta \mathcal{O}} = 0. \tag{1.4d}$$

When $\mathcal{O}$ is bosonic, a useful extension of the above result states that the vanishing of $\delta \mathbf{P}$ for all self-adjoint variations $\delta \mathcal{O}$, or alternatively, for all anti-self-adjoint variations $\delta \mathcal{O}$, still implies the vanishing of $\delta \mathbf{P}/\delta \mathcal{O}$. To prove this, split each C_n in Eq. (1.4a) into self-adjoint and anti-self-adjoint parts, $C_n = C_n^{\mathrm{sa}} + C_n^{\mathrm{asa}}$, with $C_n^{\mathrm{sa}} = C_n^{\mathrm{sa}\dagger}$ and $C_n^{\mathrm{asa}} = -C_n^{\mathrm{asa}\dagger}$. For self-adjoint $\delta \mathcal{O}$, Eq. (1.1a) implies that $\mathrm{Tr} C_n^{\mathrm{sa}} \delta \mathcal{O}$ is real, and $\mathrm{Tr} C_n^{\mathrm{asa}} \delta \mathcal{O}$ is imaginary, and so by the reasoning of Eqs. (1.4a–c), the vanishing of $\delta \mathbf{P}$ implies that both of these traces must vanish separately. Taking $\delta \mathcal{O} = C_p^{\mathrm{sa}}$ then implies the vanishing of C_p^{sa}, while taking $\delta \mathcal{O} = i C_p^{\mathrm{asa}}$ then implies the vanishing of C_p^{asa}. A similar argument, with the role of reals and imaginaries interchanged (or equivalently, with multiplication of $\mathcal{O}$ by the c-number i) applies to the case in which $\delta \mathcal{O}$ is restricted to be anti-self-adjoint.

In our applications, we shall often consider trace functionals $\mathbf{P}$ that are real, which will be true when the adjointness properties of the operators from which P is constructed imply that $P - P^\dagger$ is either zero or is an operator with identically vanishing trace. Real trace functionals $\mathbf{P}$ have the important property that when $\mathcal{O}$ is a self-adjoint bosonic operator, then $\delta \mathbf{P}/\delta \mathcal{O}$ is also self-adjoint. To prove this, we make a self-adjoint variation $\delta \mathcal{O}$, and use the reality of $\mathbf{P}$ to write

$$0 \equiv \mathrm{Im} \mathrm{Tr} \delta \mathbf{P} \propto \mathrm{Tr} \left[\frac{\delta \mathbf{P}}{\delta \mathcal{O}} \delta \mathcal{O} - (\delta \mathcal{O})^\dagger \left(\frac{\delta \mathbf{P}}{\delta \mathcal{O}} \right)^\dagger \right]$$

$$= \mathrm{Tr} \delta \mathcal{O} \left[\frac{\delta \mathbf{P}}{\delta \mathcal{O}} - \left(\frac{\delta \mathbf{P}}{\delta \mathcal{O}} \right)^\dagger \right]. \tag{1.5}$$

This implies, by the extension given in the preceding paragraph, that the anti-self-adjoint part of $\delta \mathbf{P}/\delta \mathcal{O}$ must vanish. Similarly, when $\mathbf{P}$ is real and $\delta \mathcal{O}$ is anti-self-adjoint, then $\delta \mathbf{P}/\delta \mathcal{O}$ is also anti-self-adjoint.

1.3 Lagrangian and Hamiltonian dynamics of matrix models

We can now proceed to use the apparatus just described to set up a Lagrangian and Hamiltonian dynamics for matrix models. Let $L[\{q_r\}, \{\dot{q}_r\}]$ be a Grassmann even polynomial function of the bosonic or fermionic operators $\{q_r\}$ and their time derivatives $\{\dot{q}_r\}$, which are all assumed to obey the cyclic relations of Eqs. (1.1a–c) and (1.2) under the trace. The discrete index r labels the matrix degrees of freedom for a general matrix dynamics, and in later field theory applications will be taken as a label of infinitesimal spatial boxes. Just as a classical dynamical system can have any number of degrees of freedom, the numbers n_B and n_F of bosonic and fermionic operators $\{q_r\}$ are arbitrary, and are unrelated to the dimension N of the matrices that represent these operators. From L, we form the *trace Lagrangian*

$$\mathbf{L}[\{q_r\}, \{\dot{q}_r\}] = \mathrm{Tr} L[\{q_r\}, \{\dot{q}_r\}], \tag{1.6a}$$

and the corresponding *trace action*

$$\mathbf{S} = \int dt \mathbf{L}. \tag{1.6b}$$

We shall assume that the trace action is real valued, which requires that L be self-adjoint up to a possible total time derivative and/or a possible term with vanishing trace, such as a commutator. That is, we require

$$L - L^\dagger = \frac{d}{dt}\Delta_1 + [\Delta_2, \Delta_3], \tag{1.6c}$$

with $\Delta_{1,2,3}$ arbitrary. Requiring that the trace action be stationary with respect to variations of the q_rs that preserve their bosonic or fermionic type, and using the definition of Eq. (1.3b), we get

$$0 = \delta\mathbf{S} = \int dt \mathrm{Tr} \sum_r \left(\frac{\delta\mathbf{L}}{\delta q_r}\delta q_r + \frac{\delta\mathbf{L}}{\delta\dot{q}_r}\delta\dot{q}_r \right), \tag{1.7a}$$

or after integrating by parts in the second term and discarding surface terms

$$0 = \delta\mathbf{S} = \int dt \mathrm{Tr} \sum_r \left(\frac{\delta\mathbf{L}}{\delta q_r} - \frac{d}{dt}\frac{\delta\mathbf{L}}{\delta\dot{q}_r} \right) \delta q_r. \tag{1.7b}$$

For this to hold for general same-type operator variations δq_r, the coefficient of each δq_r in Eq. (1.7b) must vanish for all t, giving the operator Euler–Lagrange equations

$$\frac{\delta\mathbf{L}}{\delta q_r} - \frac{d}{dt}\frac{\delta\mathbf{L}}{\delta\dot{q}_r} = 0. \tag{1.7c}$$

Because, by the definition of Eq. (1.3b), we have

$$\left(\frac{\delta \mathbf{L}}{\delta q_r}\right)_{ij} = \frac{\partial \mathbf{L}}{\partial (q_r)_{ji}}, \tag{1.8}$$

for each r the single Euler–Lagrange equation of Eq. (1.7c) is equivalent to the N^2 Euler–Lagrange equations obtained by regarding $\mathbf{L}$ as a function of the N^2 matrix element variables $(q_r)_{ji}$. (For future reference, we note that the identity of Eq. (1.8) still holds when $\mathbf{L}$ is replaced by a general complex valued trace functional $\mathbf{A}$.) Let us now define the momentum operator p_r conjugate to q_r by

$$p_r \equiv \frac{\delta \mathbf{L}}{\delta \dot{q}_r}, \tag{1.9a}$$

so that the Euler–Lagrange equations take the form $\delta \mathbf{L}/\delta q_r = \dot{p}_r$. Since the Lagrangian is Grassmann even, p_r is of the same bosonic or fermionic type as q_r. We can now introduce a trace Hamiltonian $\mathbf{H}$ by analogy with the usual definition

$$\mathbf{H} = \mathrm{Tr} \sum_r p_r \dot{q}_r - \mathbf{L}. \tag{1.9b}$$

In correspondence with Eq. (1.8), the matrix elements $(p_r)_{ij}$ of the momentum operator p_r just correspond to the momenta canonical to the matrix element variables $(q_r)_{ji}$. Performing general same-type operator variations, and using Eq. (1.9a) and the Euler–Lagrange equations, we find from Eq. (1.9b) that

$$\begin{aligned}
\delta \mathbf{H} &= \mathrm{Tr} \sum_r \left((\delta p_r)\dot{q}_r + p_r \delta \dot{q}_r\right) - \mathrm{Tr} \sum_r \left(\frac{\delta \mathbf{L}}{\delta q_r}\delta q_r + \frac{\delta \mathbf{L}}{\delta \dot{q}_r}\delta \dot{q}_r\right) \\
&= \mathrm{Tr} \sum_r \left((\delta p_r)\dot{q}_r - \dot{p}_r \delta q_r\right) \\
&= \mathrm{Tr} \sum_r \left(\epsilon_r \dot{q}_r \delta p_r - \dot{p}_r \delta q_r\right).
\end{aligned} \tag{1.9c}$$

Therefore the trace Hamiltonian $\mathbf{H}$ is a trace functional of the operators $\{q_r\}$ and $\{p_r\}$

$$\mathbf{H} = \mathbf{H}[\{q_r\}, \{p_r\}], \tag{1.10a}$$

with the operator derivatives

$$\frac{\delta \mathbf{H}}{\delta q_r} = -\dot{p}_r, \quad \frac{\delta \mathbf{H}}{\delta p_r} = \epsilon_r \dot{q}_r, \tag{1.10b}$$

where $\epsilon_r = 1(-1)$ according to whether q_r, p_r are bosonic (fermionic).

1.4 The generalized Poisson bracket, its properties, and applications

Letting $\mathbf{A}$ and $\mathbf{B}$ be two bosonic trace functionals of the operators $\{q_r\}$ and $\{p_r\}$, it is convenient to define the *generalized Poisson bracket*

$$\{\mathbf{A}, \mathbf{B}\} = \mathrm{Tr} \sum_r \epsilon_r \left(\frac{\delta \mathbf{A}}{\delta q_r} \frac{\delta \mathbf{B}}{\delta p_r} - \frac{\delta \mathbf{B}}{\delta q_r} \frac{\delta \mathbf{A}}{\delta p_r} \right). \tag{1.11a}$$

Then using the Hamiltonian form of the equations of motion, one readily finds that for a general bosonic trace functional $\mathbf{A}[\{q_r\}, \{p_r\}, t]$, the time derivative is given by

$$\begin{aligned}
\frac{d}{dt}\mathbf{A} &= \frac{\partial \mathbf{A}}{\partial t} + \mathrm{Tr} \sum_r \left(\frac{\delta \mathbf{A}}{\delta q_r} \dot{q}_r + \frac{\delta \mathbf{A}}{\delta p_r} \dot{p}_r \right) \\
&= \frac{\partial \mathbf{A}}{\partial t} + \mathrm{Tr} \sum_r \left(\frac{\delta \mathbf{A}}{\delta q_r} \epsilon_r \frac{\delta \mathbf{H}}{\delta p_r} - \frac{\delta \mathbf{A}}{\delta p_r} \frac{\delta \mathbf{H}}{\delta q_r} \right) \\
&= \frac{\partial \mathbf{A}}{\partial t} + \mathrm{Tr} \sum_r \epsilon_r \left(\frac{\delta \mathbf{A}}{\delta q_r} \frac{\delta \mathbf{H}}{\delta p_r} - \frac{\delta \mathbf{H}}{\delta q_r} \frac{\delta \mathbf{A}}{\delta p_r} \right) \\
&= \frac{\partial \mathbf{A}}{\partial t} + \{\mathbf{A}, \mathbf{H}\}.
\end{aligned} \tag{1.11b}$$

In particular, letting $\mathbf{A}$ be the trace Hamiltonian $\mathbf{H}$, which has no explicit time dependence when the Lagrangian has no explicit time dependence, and using the fact that the generalized Poisson bracket is antisymmetric in its arguments, it follows that the time derivative of $\mathbf{H}$ vanishes

$$\frac{d}{dt}\mathbf{H} = 0. \tag{1.12}$$

An important property of the generalized Poisson bracket is that it satisfies the Jacobi identity

$$\{\mathbf{A}, \{\mathbf{B}, \mathbf{C}\}\} + \{\mathbf{C}, \{\mathbf{A}, \mathbf{B}\}\} + \{\mathbf{B}, \{\mathbf{C}, \mathbf{A}\}\} = 0. \tag{1.13a}$$

This can be proved algebraically in a basis independent way following Adler, Bhanot, and Weckel (1994), as explained in Appendix B, and can also be proved (Adler, 1994, App. A) by inserting a complete set of intermediate states into the trace on the right of Eq. (1.11a) and using the complex valued analogs of Eq. (1.8), giving

$$\begin{aligned}
\{\mathbf{A}, \mathbf{B}\} &= \sum_{m,n,r} \epsilon_r \left[\left(\frac{\delta \mathbf{A}}{\delta q_r} \right)_{mn} \left(\frac{\delta \mathbf{B}}{\delta p_r} \right)_{nm} - \left(\frac{\delta \mathbf{B}}{\delta q_r} \right)_{mn} \left(\frac{\delta \mathbf{A}}{\delta p_r} \right)_{nm} \right] \\
&= \sum_{m,n,r} \epsilon_r \left[\frac{\partial \mathbf{A}}{\partial (q_r)_{nm}} \frac{\partial \mathbf{B}}{\partial (p_r)_{mn}} - \frac{\partial \mathbf{B}}{\partial (q_r)_{nm}} \frac{\partial \mathbf{A}}{\partial (p_r)_{mn}} \right].
\end{aligned} \tag{1.13b}$$

In the second line of Eq. (1.13b), the generalized Poisson bracket has been reexpressed as a sum of classical Poisson brackets in which the matrix elements of q_r, p_r are the classical variables (which, depending on whether r is a bosonic or fermionic index, are both either even or odd elements of a Grassmann algebra), and the Jacobi identity of Eq. (1.13a) then follows from the Jacobi identity for the classical Poisson bracket (as extended to a Grassmann algebra). As a result of the Jacobi identity, if $\mathbf{Q_1}$ and $\mathbf{Q_2}$ are two conserved charges with no explicit time dependence, that is if

$$0 = \frac{d}{dt}\mathbf{Q_1} = \{\mathbf{Q_1}, \mathbf{H}\}, \quad 0 = \frac{d}{dt}\mathbf{Q_2} = \{\mathbf{Q_2}, \mathbf{H}\}, \tag{1.13c}$$

then their generalized Poisson bracket $\{\mathbf{Q_1}, \mathbf{Q_2}\}$ also has a vanishing generalized Poisson bracket with $\mathbf{H}$, and is conserved. This has the consequence that Lie algebras of symmetries can be represented as Lie algebras of trace functionals under the generalized Poisson bracket operation.

More generally, the Jacobi identity implies that trace dynamics has an underlying symplectic geometry that is preserved by the time evolution generated by the trace Hamiltonian, in analogy with corresponding symplectic structures in classical dynamics. This is discussed in more detail in Appendix C, following Adler and Wu (1994). Although it will not play a role in the sequel, we note for completeness that if the algebra of trace functionals is extended so as to be closed under multiplication as well as addition of trace functionals, then the operator variational derivative and the generalized Poisson bracket both obey the Leibniz product rule

$$\frac{\delta(\mathbf{AB})}{\delta q_r} = \frac{\delta\mathbf{A}}{\delta q_r}\mathbf{B} + \mathbf{A}\frac{\delta\mathbf{B}}{q_r},$$

$$\frac{\delta(\mathbf{AB})}{\delta p_r} = \frac{\delta\mathbf{A}}{\delta p_r}\mathbf{B} + \mathbf{A}\frac{\delta\mathbf{B}}{p_r},$$

$$\{\mathbf{AB}, \mathbf{C}\} = \{\mathbf{A}, \mathbf{C}\}\mathbf{B} + \mathbf{A}\{\mathbf{B}, \mathbf{C}\}. \tag{1.14}$$

Hence the extended algebra of trace functionals forms a so-called Poisson algebra (see, e.g., Giulini, 2003) under the combined operations of ordinary multiplication of traces and the generalized Poisson bracket.

It will be useful at this point to introduce a compact notation for the operator phase space variables, which emphasizes the symplectic structure. Let us introduce the notation $x_1 = q_1$, $x_2 = p_1$, $x_3 = q_2$, $x_4 = p_2$, ..., $x_{2D-1} = q_D$, $x_{2D} = p_D$, where by convention we list all of the bosonic variables before all of the fermionic ones in the $2D$-dimensional phase space vector x_r, with $D = n_B + n_F$.

The generalized Poisson bracket of Eq. (1.11a) can now be rewritten as

$$\{\mathbf{A}, \mathbf{B}\} = \text{Tr} \sum_{r,s=1}^{2D} \left(\frac{\delta \mathbf{A}}{\delta x_r} \omega_{rs} \frac{\delta \mathbf{B}}{\delta x_s} \right), \tag{1.15a}$$

and the operator Hamiltonian equations of Eq. (1.10b) can be compactly rewritten as

$$\dot{x}_r = \sum_{s=1}^{2D} \omega_{rs} \frac{\delta \mathbf{H}}{\delta x_s}. \tag{1.15b}$$

The numerical matrix ω_{rs} that appears here is given by

$$\omega = \text{diag}(\Omega_B, \ldots, \Omega_B, \Omega_F, \ldots, \Omega_F), \tag{1.16a}$$

with the 2×2 bosonic and fermionic matrices Ω_B and Ω_F given respectively by

$$\Omega_B = \begin{pmatrix} 0 & 1 \\ -1 & 0 \end{pmatrix}, \quad \Omega_F = -\begin{pmatrix} 0 & 1 \\ 1 & 0 \end{pmatrix}. \tag{1.16b}$$

It is easy to verify that the matrix ω obeys the properties

$$(\omega^2)_{rs} = -\epsilon_r \delta_{rs}, \quad \omega_{sr} = -\epsilon_r \omega_{rs} = -\epsilon_s \omega_{rs},$$
$$(\omega^4)_{rs} = \delta_{rs}, \quad \sum_r \omega_{rs} \omega_{rt} = \sum_r \omega_{sr} \omega_{tr} = \delta_{st}. \tag{1.17}$$

Henceforth, as in Eq. (1.17), we shall not explicitly indicate the range of the summation indices; the index r on q_r, p_r will be understood to have an upper summation limit of D, while the index r on x_r will be understood to have an upper limit of $2D$.

Using this compact notation one can formally integrate the trace dynamics equations of motion. Let j_r be a constant source matrix of the same bosonic or fermionic type as x_r, and let us define

$$\mathbf{X}_r = \text{Tr} j_r x_r, \tag{1.18a}$$

so that

$$\frac{\delta \mathbf{X}_r}{\delta x_u} = \delta_{ru} j_r. \tag{1.18b}$$

Then the Hamiltonian equations of motion of Eq. (1.15b) can be rewritten, following Adler and Horwitz (1996), as

$$\dot{\mathbf{X}}_r = \text{Tr} j_r \dot{x}_r = \text{Tr} \sum_u \delta_{ru} j_r \dot{x}_u$$

$$= \text{Tr} \sum_{s,u} \frac{\delta \mathbf{X}_r}{\delta x_u} \omega_{us} \frac{\delta \mathbf{H}}{\delta x_s} = \{\mathbf{X}_r, \mathbf{H}\} = -\{\mathbf{H}, \mathbf{X}_r\}, \tag{1.18c}$$

which expresses $\dot{\mathbf{X}}_r$ as a generalized Poisson bracket with the trace Hamiltonian. We can now formally integrate the equation of motion for $\mathbf{X}_r(t)$ by writing

$$\mathbf{X}_r(t) = \exp(-\{\mathbf{H}, \ldots\}t)\mathbf{X}_r(0)\exp(\{\mathbf{H}, \ldots\}t)$$

$$= \mathbf{X}_r(0) - t\{\mathbf{H}, \mathbf{X}_r(0)\} + \frac{1}{2}t^2\{\mathbf{H}, \{\mathbf{H}, \mathbf{X}_r(0)\}\}$$

$$-\frac{1}{6}t^3\{\mathbf{H}, \{\mathbf{H}, \{\mathbf{H}, \mathbf{X}_r(0)\}\}\} + \ldots \tag{1.19}$$

1.5 Trace dynamics contrasted with unitary Heisenberg picture dynamics

In general, the matrix dynamics specified by Eqs. (1.15b) and (1.19) is not unitary, in other words, Eq. (1.19) is *not* equivalent to an evolution of the form

$$x_r(t) = U^\dagger(t)x_r(0)U(t), \tag{1.20a}$$

for some unitary $U(t)$. Expressed in generator form, by writing $U(t)$ as $U(t) = \exp(-iGt)$, the evolution of Eq. (1.15b) is *not* equivalent to a Heisenberg picture time evolution (with $\hbar = 1$)

$$\dot{x}_r(t) = i[G, x_r(t)]. \tag{1.20b}$$

As a concrete example, consider the case with two degrees of freedom and the trace Hamiltonian $\mathbf{H} = \mathrm{Tr}\,i\{q_1, p_2\}[q_1, p_1]$, with p_1, p_2, and q_1 bosonic operators. Then from Eq. (1.10b) we compute

$$\dot{q}_1 = i[\{q_1, p_2\}, q_1],$$
$$\dot{q}_2 = i\{q_1, [q_1, p_1]\},$$
$$\dot{p}_1 = -i\left(\{p_2, [q_1, p_1]\} + [p_1, \{q_1, p_2\}]\right), \tag{1.20c}$$
$$\dot{p}_2 = 0,$$

which clearly cannot be represented in the unitary evolution form given in Eq. (1.20b) for any choice of generator G. This statement holds true even when Eq. (1.20c) is simplified by assuming the canonical algebra $[q_r, p_s] = i\delta_{rs}$, $[q_r, q_s] = [p_r, p_s] = 0$ (which involves an extension of the algebra of dynamical variables to ones that are not trace class), so that the equations of motion read

$$\dot{q}_1 = 0,$$
$$\dot{q}_2 = -2q_1, \tag{1.20d}$$
$$\dot{p}_1 = 0,$$
$$\dot{p}_2 = 0.$$

These equations cannot be represented, for any choice of G, as a Heisenberg evolution $\dot{x} = i[G, x]$ for all $x = q_1, q_2, p_1, p_2$, since the second line in Eq. (1.20d) implies, assuming the canonical algebra, that G must contain a term proportional to $p_2 q_1$, whereas the third line in Eq. (1.20d) implies, together with the canonical algebra, that G has no dependence on q_1! Equations (1.20d) can, however, at one instant of time be represented as different Heisenberg evolutions for the two canonical pairs q_1, p_1 and q_2, p_2

$$
\begin{aligned}
\dot{q}_1 &= i[G_1, q_1], \\
\dot{q}_2 &= i[G_2, q_2], \\
\dot{p}_1 &= i[G_1, p_1], \\
\dot{p}_2 &= i[G_2, p_2],
\end{aligned}
\tag{1.20e}
$$

with $G_1 = 0$ and $G_2 = -\{q_1, p_2\}$, a statement that we shall see can be extended to the general case. The restriction of Eq. (1.20e) to one instant of time is required because, in this example, although the first time derivatives of the relations $[q_1, p_1] = i$, $[q_2, p_2] = i$, $[q_1, q_2] = 0$, and $[p_1, p_2] = 0$ are consistent with the equations of motion of Eqs. (1.20d,e), the first time derivative of the relation $[p_1, q_2] = 0$ is not, since

$$
[\dot{p}_1, q_2] + [p_1, \dot{q}_2] = -2[p_1, q_1] = 2i \neq 0.
\tag{1.20f}
$$

In general, as we shall remark again below, multiple Heisenberg-like evolution does not preserve an initially assumed canonical algebra, and so can be used to represent the trace dynamics equations of motion only at one instant of time.

There is, however, a special case, discussed in Adler and Millard (1996), in which the trace dynamics and the unitary Heisenberg picture evolutions coincide. Let us consider a special class of operator Hamiltonians called *Weyl-ordered* Hamiltonians, in which the bosonic operators are all totally symmetrized with respect to one another and to the fermionic operators, and in which the fermionic operators are totally antisymmetrized with respect to one another. (Note that in the conventional quantum mechanical application of Weyl ordering, in which operators such as q_1 and p_2 commute according to the canonical algebra, their products do not need to be symmetrized; in trace dynamics, since no a priori commutativity properties are assumed, Weyl ordering requires the symmetrization/antisymmetrization of all operator products.) Clearly, the most general Weyl-ordered Hamiltonian which is a polynomial in the operator phase space variables $\{x_r\}$ will be a sum of terms, which may be of different degrees, each obtained by Weyl ordering a distinct monomial in the phase space variables. The contribution of all such monomials of degree n may be simply represented by a generating function G_n constructed as follows. Let σ_r, $r = 1, \ldots, 2D$ be a set of parameters which are real numbers when $\epsilon_r = 1$ and which are real Grassmann numbers, which anticommute with each other and with all of the fermionic phase space variables, when $\epsilon_r = -1$. Then if we form

$$G_n = g^n, \quad g = \sum_s \sigma_s x_s, \tag{1.21a}$$

the coefficient of each distinct monomial in the parameters σ_r will be a distinct Weyl-ordered polynomial of degree n in the phase space variables $\{x_r\}$. Corresponding to the operator generating function G_n, we define a trace functional generating function

$$\mathbf{G}_n = \mathrm{Tr} G_n. \tag{1.21b}$$

The part of $\mathbf{G}_n$ which is even in the Grassmann parameters is then a generating function for all non-vanishing trace functionals that correspond to the bosonic Weyl-ordered monomials generated by G_n.

Let us now compare the trace dynamics equations of motion produced by $\mathbf{G}_n$ for general operators $\{x_r\}$, with the corresponding Heisenberg picture equations of motion produced by G_n when the phase space variables $\{x_r\}$ are assumed to obey the standard canonical algebra of quantum mechanics. In our compact phase space notation, this algebra takes the form

$$x_r x_s - \epsilon_r x_s x_r = i \epsilon_r \omega_{rs},$$
$$[x_r, i] = 0, \tag{1.22a}$$

where we adopt the convention that if only *one* of x_r, x_s is bosonic, it is taken to be the operator x_r; alternatively, we can rewrite the first line of Eq. (1.22a) with no restrictions on the indices r, s by including a factor σ_s, giving

$$[x_r, \sigma_s x_s] = i \omega_{rs} \sigma_s. \tag{1.22b}$$

As noted above, imposition of the canonical algebra involves an extension outside the algebra of trace class matrix dynamical variables, since taking the trace of Eq. (1.22b) leads to an inconsistency if one simultaneously assumes the validity of cyclic permutation under the trace.

Applying the equations of motion of Eq. (1.15b) with $\mathbf{G}_n$ playing the role of the trace Hamiltonian, we get

$$\dot{x}_r = \sum_s \omega_{rs} \frac{\delta \mathbf{G}_n}{\delta x_s} = \sum_s \omega_{rs} n g^{n-1} \sigma_s. \tag{1.23a}$$

On the other hand, from the canonical algebra of Eq. (1.22b) we find, for both bosonic and fermionic x_r, that

$$[x_r, g] = i \sum_s \omega_{rs} \sigma_s, \qquad (1.23b)$$

which in turn implies that

$$[x_r, G_n] = n g^{n-1} i \sum_s \omega_{rs} \sigma_s. \qquad (1.23c)$$

But the Heisenberg picture equations of motion for the phase space variables, taking G_n as the operator Hamiltonian, are

$$\dot{x}_r = i[G_n, x_r], \qquad (1.23d)$$

which substituting Eq. (1.23c) becomes

$$\dot{x}_r = \sum_s \omega_{rs} n g^{n-1} \sigma_s, \qquad (1.23e)$$

in agreement with Eq. (1.23a). We can now sum over all generating function contributions G_n weighted by c-number coefficients to obtain a general Weyl-ordered Hamiltonian H, which has a corresponding trace Hamiltonian $\mathbf{H} = \mathrm{Tr} H$, which respectively generate the Heisenberg picture equation of motion

$$\dot{x}_r = i[H, x_r] \qquad (1.24a)$$

and the corresponding trace dynamics equation of motion of Eq. (1.15b).

Thus, for Weyl-ordered Hamiltonians formed with c-number coefficients, we conclude that the trace dynamics equations of motion generated by $\mathbf{H}$ agree with the Heisenberg picture equations of motion generated by H, on an initial time slice on which the phase space variables are canonical. It is also evident that on this time slice

$$[H, i] = 0. \qquad (1.24b)$$

But since Eq. (1.24b) guarantees that the Heisenberg picture equations of motion preserve the canonical algebra on the next time slice, integrating forward in time step by step then implies that trace dynamics agrees with Heisenberg picture dynamics at all subsequent times, and therefore can be extended to a unitary dynamics in this case.

When H is not Weyl ordered, as we have seen above, one can give explicit examples in which the trace dynamics equations of motion do not correspond to a unitary evolution, even when extended to the canonical algebra. The example of Eq. (1.20c) required the use of two pairs of canonical variables $q_{1,2}$, $p_{1,2}$; using our Weyl-ordering result, we shall now show that any trace dynamics for a single pair of bosonic variables q, p, when extended to the canonical algebra $[q, p] = i$, always can be represented as a unitary Heisenberg evolution. We shall proceed by induction, and assume that the result has been proved for any trace dynamics generated by a trace Hamiltonian $\mathbf{H} = \mathrm{Tr}H$ of degree n or less in p and q. Now consider a trace Hamiltonian of degree $n + 1$, and suppose that the result has been proved for those degree $n + 1$ trace Hamiltonians that can be reduced to Weyl-ordered form by at most k interchanges of p and q. This hypothesis is true for $k = 0$, i.e., the case in which we start from a degree $n + 1$ Weyl-ordered Hamiltonian. Consider now a trace Hamiltonian $\mathbf{H}_{n+1,k+1}$ which is of degree $n + 1$ and which requires $k + 1$ interchanges of p and q to reduce it to Weyl-ordered form. Doing the first of these $k + 1$ interchanges of p and q we have

$$\mathbf{H}_{n+1,k+1} = \mathbf{H}_{n+1,k} + i\mathrm{Tr}X[q, p], \tag{1.25a}$$

with X a self-adjoint polynomial of degree $n - 1$, where we have used the fact that irrespective of the position of the q and p that are being interchanged, the resulting commutator can always be cyclically permuted to the right. Varying Eq. (1.25a) we have

$$\delta\mathbf{H}_{n+1,k+1} = \delta\mathbf{H}_{n+1,k} + i\mathrm{Tr}(\delta X[q, p] + X[\delta q, p] + X[q, \delta p]). \tag{1.25b}$$

Now that we have taken variations, we can simplify the first term on the right-hand side by assuming the canonical algebra, so that after using cyclic invariance to rearrange the second and third terms on the right we get

$$\delta\mathbf{H}_{n+1,k+1} = \delta\mathbf{H}_{n+1,k} - \delta X + i\mathrm{Tr}([p, X]\delta q + [X, q]\delta p). \tag{1.25c}$$

Thus Eq. (1.25c) implies that

$$\dot{p} = -\frac{\delta\mathbf{H}_{n+1,k+1}}{\delta q} = -\frac{\delta\mathbf{H}_{n+1,k}}{\delta q} + \frac{\delta X}{\delta q} + i[X, p],$$

$$\dot{q} = \frac{\delta\mathbf{H}_{n+1,k+1}}{\delta p} = \frac{\delta\mathbf{H}_{n+1,k}}{\delta p} - \frac{\delta X}{\delta p} + i[X, q]. \tag{1.25d}$$

Now let us use the induction hypothesis, which states that the trace dynamics equations generated by both $\mathbf{H}_{n+1,k}$ and $\mathbf{X}$ simplify, over the canonical algebra, to Heisenberg picture equations of motion with generators $G_{n+1,k}$ and G_X respectively. Thus we have

$$\dot{p} = i[G_{n+1,k}, p] - i[G_X, p] + i[X, p] = i[G_{n+1,k+1}, p],$$
$$\dot{q} = i[G_{n+1,k}, q] - i[G_X, q] + i[X, q] = i[G_{n+1,k+1}, q], \tag{1.26a}$$

where we have defined

$$G_{n+1,k+1} = G_{n+1,k} - G_X + X, \tag{1.26b}$$

completing the induction. Therefore any trace dynamics in a single pair of bosonic dynamical variables q, p simplifies, when extended over the canonical algebra, to a unitary Heisenberg dynamics.

When more than one canonical pair of bosonic or fermionic dynamical variables is present, the inductive argument just given generalizes in the following way. The key step of Eq. (1.25a) now contains a sum of commutators of each bosonic variable with all other bosonic and fermionic variables, and of anticommutators of each fermionic variable with all other fermionic variables. The cyclic rearrangements and inductive strategy used above now can be used to prove that when restricted to the canonical algebra, the general trace dynamics equations of motion can be expressed at one instant of time as a multiple Heisenberg-like evolution with different generators linking each distinct pair of operators. For example, considering for simplicity just the case in which only bosonic dynamical variables are present, Eq. (1.25a) generalizes to

$$\mathbf{H}_{n+1,k+1} = \mathbf{H}_{n+1,k} + i \operatorname{Tr} \sum_{rs} (X^1_{[rs]}[q_r, q_s] + X^2_{[rs]}[p_r, p_s] + X^3_{rs}[q_r, p_s]).$$

(1.27a)

Here the Xs are independent bosonic generators for each pair of index labels, and we have followed the usual convention of denoting antisymmetry in subscripted indices by [ ]. The inductive argument then shows that when simplified over the canonical algebra, at one instant of time the trace dynamics equations of motion take the form of a multiple Heisenberg-like evolution

$$\dot{q}_r = i \sum_s \left([G^3_{sr}, q_s] + 2[G^2_{[sr]}, p_s]\right),$$

$$\dot{p}_r = i \sum_s \left([G^3_{rs}, p_s] + 2[G^1_{[rs]}, q_s]\right),$$

(1.27b)

with the generators G obtained by combining the generators X with the generators $G_{n+1,k}$ and G_X furnished by the inductive hypothesis. In the most general case, the generators appearing in Eq. (1.27b) are all distinct, but in special cases some can be zero or identical to others. When fermionic as well as bosonic variables are present, there will be additional terms in Eq. (1.27a) involving commutators of bosonic variables with fermionic variables, and anticommutators of fermionic variables with one another, each with an independent operator coefficient, which give rise to corresponding additional terms in Eq. (1.27b).

The restriction to one instant of time is needed because the evolution of Eq. (1.27b) does not preserve the structure of the canonical algebra (as is easily seen by application of the Jacobi identity for commutators), and so the trace dynamics equations of motion, for an asymmetric Hamiltonian in the many variable case, cannot be replaced by a canonical evolution even with the complicated generator structure given in Eq. (1.27b). However, because of the indicated index antisymmetry of $G^1_{[rs]}$ and $G^2_{[rs]}$, when only one bosonic canonical pair is present the evolution of Eq. (1.27b) reduces to the much simpler Heisenberg evolution of Eq. (1.26a), and so we recover our previous result.

The result of Eq. (1.27b) shows that when two or more canonical pairs of dynamical variables are present, in general Weyl ordering of the Hamiltonian is needed for the trace dynamics equations of motion to simplify, when extended over the canonical algebra, to unitary Heisenberg ones. When the trace dynamics equations of motion for the generic case of an unsymmetrized Hamiltonian are extended to the canonical algebra, the tightly constrained Heisenberg evolution found in the Weyl-ordered case, in which each dynamical variable evolves with the same Hamiltonian generator, "fragments" into independent evolutions for each dynamical variable. This phenomenon is relevant because non-Weyl-ordered Hamiltonians appear in models of physical interest, such as the supersymmetric Yang–Mills models described in Sections 3.2 and 3.3, and the non-supersymmetric but operator gauge invariant models discussed in Appendix E, all of which involve commutator terms in the construction of their Hamiltonians. More generally, the fact that the trace dynamics equations of motion are not a unitary Heisenberg evolution plays a role in Chapter 6, where we argue that the leading corrections to an approximation of a unitary Heisenberg evolution take the form of rapidly fluctuating terms in the Schrödinger equation, that provide the theoretical basis for stochastic localization models for state vector reduction.

To conclude this discussion, we emphasize that in the remainder of this book we will *not* assume that the matrix variables obey canonical commutation/anticommutation relations. They are instead assumed to be completely general trace class matrix operators with no special commutativity/anticommutativity properties. An effective canonical algebra will be seen to hold as an approximation only for averages of the matrix variables over a statistical mechanical canonical ensemble. The discussion of this section then shows that, for a trace dynamics generated from a Weyl-ordered Hamiltonian, this effective canonical algebra and Heisenberg dynamics gives equations of motion that agree with those arising from the underlying trace dynamics.

2

Additional generic conserved quantities

We have seen in Chapter 1 that the trace Hamiltonian **H** is always a conserved quantity in the dynamics of matrix models. In this chapter we introduce two structural restrictions on the form of the trace Hamiltonian (or Lagrangian), which lead to two further generic conserved quantities. The first conserved quantity, discussed in Section 2.1, is a trace quantity **N** analogous to the fermion number operator in field theory, and the second, introduced in Section 2.2, is an operator $\tilde{C}$ that is reminiscent of the canonical commutator/anticommutator structure of field theory. These conserved quantities play a central role in the statistical mechanical analysis of Chapters 4 and 5, where we shall see that when the low energy dynamics is effectively dominated by $\tilde{C}$, with the trace quantities **H** and **N** effectively decoupled, then the statistical thermodynamics of matrix models has the structure of quantum field theory.

When the index r labeling the dynamical variables is a spatial box label, and when the corresponding trace Lagrangian is Poincaré invariant, there will be additional conserved quantities that play the role of trace generators of the Poincaré group under the generalized Poisson bracket introduced in Section 1.4. These conserved quantities are all charges associated with corresponding conserved currents, as discussed in Section 2.3. To illustrate the general trace dynamics formalism in action, and to give examples of all of the conserved quantities introduced in this Chapter, in Section 2.4 we analyze in detail the simple trace dynamics model defined by coupling a Dirac fermion to a scalar Klein–Gordon field. Finally, in Section 2.5 we discuss the symmetry properties of the conserved quantities under the interchange of fermionic canonical coordinates and momenta.

2.1 The trace "fermion number" N

Although we shall allow the trace Hamiltonian to have arbitrary polynomial dependences on the bosonic variables, let us for the moment restrict the fermionic structure to have the bilinear form found in all renormalizable quantum field theory models, by taking **H** to have the form

$$\mathbf{H} = \mathrm{Tr}H = \mathrm{Tr} \sum_{r,s \in F} (p_r q_s B_{1rs} + p_r B_{2rs} q_s) + \text{purely bosonic.} \qquad (2.1a)$$

Here the notation $\in F$ indicates a sum over only the fermionic operator phase space variables, and $B_{1,2}$ are general polynomials in the bosonic variables. (Requiring self-adjointness of H places restrictions on the form of B_{1rs} and B_{2rs} that are given in Eq. (2.4f) below, but are not needed for the conservation argument that follows.) From Eq. (2.1a) and the Hamilton equations of Eq. (1.10b), we have for fermionic r, s

$$\dot{p}_s = -\frac{\delta \mathbf{H}}{\delta q_s} = -\sum_{r \in F} (B_{1rs} p_r + p_r B_{2rs}),$$

$$\dot{q}_r = -\frac{\delta \mathbf{H}}{\delta p_r} = \sum_{s \in F} (q_s B_{1rs} + B_{2rs} q_s). \qquad (2.1b)$$

Let us now define the trace quantity **N** by

$$\mathbf{N} = \frac{1}{2} i \mathrm{Tr} \sum_{r \in F} [q_r, p_r] = i \mathrm{Tr} \sum_{r \in F} q_r p_r = -i \mathrm{Tr} \sum_{r \in F} p_r q_r. \qquad (2.2a)$$

Then for the time derivative of **N** we have, from the second of the three equivalent forms of **N**

$$\dot{\mathbf{N}} = i \mathrm{Tr} \sum_{r \in F} (\dot{q}_r p_r + q_r \dot{p}_r), \qquad (2.2b)$$

which on substituting the fermion equations of motion of Eq. (2.1b) becomes

$$\dot{\mathbf{N}} = i \mathrm{Tr} \sum_{r,s \in F} [B_{2rs}, q_s p_r] = 0. \qquad (2.3)$$

Thus, **N** is a conserved trace quantity when the trace Hamiltonian has the bilinear fermionic structure of Eq. (2.1a). Inverting the Legendre transformation of Eq. (1.9b), the corresponding trace Lagrangian is

$$\mathbf{L} = \mathrm{Tr}L = \mathrm{Tr} \sum_{r \in F} p_r \dot{q}_r - \mathrm{Tr} \sum_{r,s \in F} (p_r q_s B_{1rs} + p_r B_{2rs} q_s) + \text{purely bosonic.}$$

$$(2.4a)$$

In order for the kinetic part of L to be self-adjoint up to a total time derivative, we assign adjointness properties of the fermionic variables according to

$$p_r = q_r^\dagger, \qquad (2.4b)$$

which gives

$$(p_r \dot{q}_r)^\dagger = (q_r^\dagger \dot{q}_r)^\dagger = -\dot{q}_r^\dagger q_r$$

$$= q_r^\dagger \dot{q}_r - \frac{d}{dt}(q_r^\dagger q_r) = p_r \dot{q}_r + \text{total time derivative}, \qquad (2.4c)$$

as needed. (A more general construction of the fermionic kinetic Lagrangian, and a correspondingly more general assignment of adjointness properties of the fermionic variables, will be taken up at the end of Section 2.2.) Substituting Eq. (2.4b) into Eq. (2.4a), the trace Lagrangian takes the form

$$\mathbf{L} = \text{Tr} \sum_{r \in F} q_r^\dagger \dot{q}_r - \text{Tr} \sum_{r,s \in F} (q_r^\dagger q_s B_{1rs} + q_r^\dagger B_{2rs} q_s) + \text{purely bosonic}. \qquad (2.4d)$$

Correspondingly, substituting Eq. (2.4b) into Eq. (2.2a) for **N** we get

$$\mathbf{N} = \frac{1}{2} i \text{Tr} \sum_{r \in F} [q_r, q_r^\dagger] = i \text{Tr} \sum_{r \in F} q_r q_r^\dagger = -i \text{Tr} \sum_{r \in F} q_r^\dagger q_r, \qquad (2.4e)$$

showing that, since **N** is the trace of a self-adjoint quantity, it is real when the fermionic adjointness properties are assigned as in Eq. (2.4b). From Eq. (2.4d), we see that with the fermionic adjointness assignment of Eq. (2.4b), self-adjointness of L and H requires that

$$B_{1rs} = -B_{1sr}^\dagger, \quad B_{2rs} = -B_{2sr}^\dagger, \quad \text{all } r, s. \qquad (2.4f)$$

The resemblance of **N** to a fermion number operator suggests that it will be conserved even when **H** is not bilinear, as long as each monomial in **H** has equal numbers of fermionic operators p_r and q_s, with any values of the mode indices r, s. This is indeed the case, and can be seen as follows. Let $\mathbf{H}_{n_q,n_p}$ be a monomial term in **H** containing exactly n_q factors of fermionic qs, and n_p factors of fermionic ps, with any values of the indices $r, \ldots$ labeling fermionic degrees of freedom. Then by a simple counting argument (an application of Euler's theorem for homogeneous functions) we have

$$\text{Tr} \sum_{r \in F} \frac{\delta \mathbf{H}_{n_q,n_p}}{\delta q_r} q_r = n_q \mathbf{H}_{n_q,n_p},$$

$$\text{Tr} \sum_{r \in F} \frac{\delta \mathbf{H}_{n_q,n_p}}{\delta p_r} p_r = n_p \mathbf{H}_{n_q,n_p}. \qquad (2.5a)$$

Hence denoting by $\dot{\mathbf{N}}_{n_q,n_p}$ the contribution of $\mathbf{H}_{n_q,n_p}$ to $\dot{\mathbf{N}} = i\,\mathrm{Tr}\sum_{r\in F}(\dot{q}_r\,p_r + q_r\,\dot{p}_r)$, we have by use of Eq. (1.10b)

$$
\begin{aligned}
\dot{\mathbf{N}}_{n_q,n_p} &= -i\,\mathrm{Tr}\sum_{r\in F}\left[\frac{\delta\mathbf{H}_{n_q,n_p}}{\delta p_r}p_r + q_r\frac{\delta\mathbf{H}_{n_q,n_p}}{\delta q_r}\right]\\
&= -i\,\mathrm{Tr}\sum_{r\in F}\left[\frac{\delta\mathbf{H}_{n_q,n_p}}{\delta p_r}p_r - \frac{\delta\mathbf{H}_{n_q,n_p}}{\delta q_r}q_r\right]\\
&= -i(n_p - n_q)\mathbf{H}_{n_q,n_p}.
\end{aligned}
\tag{2.5b}
$$

Hence if $\mathbf{H}$ is constructed solely from monomials which have equal numbers of fermionic qs and ps, so that $n_q = n_p$ for all monomial terms in $\mathbf{H}$, then the trace quantity $\mathbf{N}$ remains conserved. This gives the most general structural restriction on $\mathbf{H}$ leading to conservation of $\mathbf{N}$.

An alternative derivation of the conservation of $\mathbf{N}$ in the general case shows its role as a Noether charge. When the numbers of fermionic operators p_r and q_s are matched in every monomial, the trace Lagrangian

$$
\mathbf{L} = \mathrm{Tr}\sum_{r\in F}q_r^\dagger\dot{q}_r + \text{terms with no time derivatives of fermions}
\tag{2.5c}
$$

will be invariant under the substitutions $q_r \to \exp(i\alpha)q_r$ and $q_s^\dagger \to \exp(-i\alpha)q_s^\dagger$ with α a real constant. When α is time dependent, this substitution results, to first order in α, in a shift in the trace Lagrangian

$$
\mathbf{L} \to \mathbf{L} + i\dot{\alpha}\mathrm{Tr}\sum_{r\in F}q_r^\dagger q_r = \mathbf{L} - \dot{\alpha}\mathbf{N}.
\tag{2.5d}
$$

The standard Noether's theorem argument (an exposition of which, in the context of continuum spacetime models, will be given below in Section 2.3) then tells us that the coefficient of $\dot{\alpha}$ in Eq. (2.5d) is a conserved charge, which again shows that $\mathbf{N}$ is a constant of the motion.

2.2 The conserved operator $\tilde{C}$

As a second structural specialization, let us restrict the class of matrix models under consideration to those in which the *only* non-commuting matrix quantities are the Lagrangian dynamical variables $q_r, \dot{q}_s$, or their Hamiltonian equivalents q_r, p_s, for general index values r, s. In other words, we shall assume that the trace Lagrangian and Hamiltonian are constructed from the dynamical variables using only *c*-number complex coefficients, excluding the more general case in which fixed matrix coefficients are used. With this restriction, we shall show that there is a generic conserved operator

$$\tilde{C} \equiv \sum_{r \in B} [q_r, p_r] - \sum_{r \in F} \{q_r, p_r\} = \sum_r (\epsilon_r q_r p_r - p_r q_r) = \sum_{r,s} x_r \omega_{rs} x_s, \quad (2.6)$$

with the notation $\in B$, $\in F$ denoting respectively sums over bosonic and fermionic operator phase space variables. The existence of the conserved quantity $\tilde{C}$ was first discovered by Millard (personal communication, 1995 and thesis, 1997) under the more restrictive assumption of a bosonic theory with a Weyl-ordered (i.e., symmetrized) Hamiltonian, but was soon seen to hold (Adler and Millard, 1996, and with a Noether formulation Adler and Kempf, 1998) under the less restrictive conditions assumed here.

When the Lagrangian and Hamiltonian are constructed using only c-number fixed coefficients, there is a bosonic global unitary invariance which preserves the adjointness properties of the dynamical variables (in the sense that when we set $y_r = U^\dagger x_r U$, then $y_r^\dagger = U^\dagger x_r^\dagger U$.) That is, if there are no fixed matrix coefficients, then the trace Lagrangian obeys

$$\mathbf{L}[\{U^\dagger q_r U\}, \{U^\dagger \dot{q}_r U\}] = \mathbf{L}[\{q_r\}, \{\dot{q}_r\}], \quad (2.7a)$$

and the trace Hamiltonian correspondingly obeys

$$\mathbf{H}[\{U^\dagger q_r U\}, \{U^\dagger p_r U\}] = \mathbf{H}[\{q_r\}, \{p_r\}], \quad (2.7b)$$

with U a constant unitary $N \times N$ matrix. Let us now find the conserved charge corresponding to this global unitary invariance. Setting $U = \exp \Lambda$, with Λ an anti-self-adjoint bosonic generator matrix, and expanding to first order in Λ, Eq. (2.7b) implies that

$$\mathbf{H}[\{q_r - [\Lambda, q_r]\}, \{p_r - [\Lambda, p_r]\}] = \mathbf{H}[\{q_r\}, \{p_r\}]. \quad (2.8a)$$

But applying the definition of the variation of a trace functional given in Eq. (1.3b), Eq. (2.8a) becomes

$$\mathrm{Tr} \sum_r \left(-\frac{\delta \mathbf{H}}{\delta q_r} [\Lambda, q_r] - \frac{\delta \mathbf{H}}{\delta p_r} [\Lambda, p_r] \right) = 0, \quad (2.8b)$$

which by use of the cyclic identities of Eqs. (1.2) yields

$$\mathrm{Tr} \Lambda \sum_r \left(\frac{\delta \mathbf{H}}{\delta q_r} q_r - \epsilon_r q_r \frac{\delta \mathbf{H}}{\delta q_r} + \frac{\delta \mathbf{H}}{\delta p_r} p_r - \epsilon_r p_r \frac{\delta \mathbf{H}}{\delta p_r} \right) = 0. \quad (2.8c)$$

Since the generator Λ is an arbitrary anti-self-adjoint $N \times N$ matrix, the matrix that multiplies it in Eq. (2.8c) must vanish, giving the matrix identity

$$\sum_r \left(\frac{\delta \mathbf{H}}{\delta q_r} q_r - \epsilon_r q_r \frac{\delta \mathbf{H}}{\delta q_r} + \frac{\delta \mathbf{H}}{\delta p_r} p_r - \epsilon_r p_r \frac{\delta \mathbf{H}}{\delta p_r} \right) = 0. \quad (2.9a)$$

But now substituting the Hamilton equations of Eq. (1.10b), Eq. (2.9a) takes the form

$$0 = \sum_r (-\dot{p}_r q_r + \epsilon_r q_r \dot{p}_r + \epsilon_r \dot{q}_r p_r - p_r \dot{q}_r)$$

$$= \frac{d}{dt} \sum_r (-p_r q_r + \epsilon_r q_r p_r)$$

$$= \frac{d}{dt} \left(\sum_{r \in B} [q_r, p_r] - \sum_{r \in F} \{q_r, p_r\} \right), \tag{2.9b}$$

completing the demonstration of the conservation of $\tilde{C}$. An analogous demonstration can be given starting from global unitary invariance of the trace Lagrangian **L**. In place of Eq. (2.8b) one finds

$$\text{Tr} \sum_r \left(-\frac{\delta \mathbf{L}}{\delta q_r} [\Lambda, q_r] - \frac{\delta \mathbf{L}}{\delta \dot{q}_r} [\Lambda, \dot{q}_r] \right) = 0, \tag{2.9c}$$

and in place of Eq. (2.9a) one has

$$\sum_r \left(\frac{\delta \mathbf{L}}{\delta q_r} q_r - \epsilon_r q_r \frac{\delta \mathbf{L}}{\delta q_r} + \frac{\delta \mathbf{L}}{\delta \dot{q}_r} \dot{q}_r - \epsilon_r \dot{q}_r \frac{\delta \mathbf{L}}{\delta \dot{q}_r} \right) = 0. \tag{2.9d}$$

Substituting the Euler–Lagrange equations of Eq. (1.7c), together with the definition of the canonical momentum from Eq. (1.9a), then leads again to Eq. (2.9b).

In Eqs. (2.7a,b) we have assumed that U is a time-independent unitary matrix. When U and Λ are time dependent, under the substitution $q_r \to q_r - [\Lambda, q_r]$ the time derivative of q_r transforms according to $\dot{q}_r \to \dot{q}_r - [\Lambda, \dot{q}_r] - [\dot{\Lambda}, q_r]$, and the trace Lagrangian changes, to first order in Λ, by

$$\mathbf{L} \to \mathbf{L} - \text{Tr} \sum_r \frac{\delta \mathbf{L}}{\delta \dot{q}_r} [\dot{\Lambda}, q_r]. \tag{2.9e}$$

Substituting Eq. (1.9a) and using the cyclic identities of Eqs. (1.1a,b) and (1.2), the added term can be rewritten as

$$-\text{Tr} \sum_r p_r (\dot{\Lambda} q_r - q_r \dot{\Lambda}) = -\text{Tr}\dot{\Lambda} \sum_r (\epsilon_r q_r p_r - p_r q_r) = -\text{Tr}\dot{\Lambda}\tilde{C}. \tag{2.9f}$$

Application of the standard Noether's theorem argument then tells us that the coefficient of $\dot{\Lambda}$ in Eq. (2.9f) is a conserved charge, which exhibits the role of the conserved quantity $\tilde{C}$ as the conserved Noether charge associated with global unitary invariance.

Assigning adjointness properties to the fermionic variables as in Eq. (2.4b), and taking the bosonic variables q_r to be self-adjoint (or anti-self-adjoint), which for a real trace Lagrangian implies that the corresponding bosonic p_r are respectively self-adjoint (or anti-self-adjoint), we see that the conserved operator $\tilde{C}$ is anti-self-adjoint

$$\tilde{C} = -\tilde{C}^\dagger. \tag{2.10a}$$

(A more general adjointness structure for $\tilde{C}$, corresponding to an alternative assignment of fermion adjointness properties, will be discussed shortly.) Also, from the bilinear cyclic identities of Eqs. (1.1a,b), we see that $\tilde{C}$ is traceless

$$\text{Tr}\tilde{C} = 0. \tag{2.10b}$$

Corresponding to the fact that $\tilde{C}$ is the conserved Noether charge in any matrix model with a global unitary invariance, it is easy to see (Adler and Horwitz, 1996; Adler and Millard, 1996) that $\tilde{C}$ can be used to construct the generator of global unitary transformations of the Hilbert space basis. Consider the trace functional

$$\mathbf{G}_\Lambda = \text{Tr}\Lambda\tilde{C}, \tag{2.11a}$$

with Λ a fixed bosonic anti-self-adjoint operator, which can be rewritten, using cyclic invariance of the trace, as

$$\mathbf{G}_\Lambda = -\text{Tr}\sum_r [\Lambda, p_r]q_r = \text{Tr}\sum_r p_r[\Lambda, q_r]. \tag{2.11b}$$

Hence for the variations of p_r and q_r induced by using $\mathbf{G}_\Lambda$ as canonical generator, which by definition (see Eq. (2.13a) below) have a structure analogous to the Hamilton equations of Eq. (1.10b), we get

$$\delta p_r \equiv -\frac{\delta \mathbf{G}_\Lambda}{\delta q_r} = [\Lambda, p_r], \qquad \delta q_r \equiv \epsilon_r \frac{\delta \mathbf{G}_\Lambda}{\delta p_r} = [\Lambda, q_r]. \tag{2.11c}$$

Comparing with Eqs. (2.7b) and (2.8a), we see that these have just the form of an infinitesimal global unitary transformation.

The generalized Poisson bracket of the trace generators of two infinitesimal global unitary transformations $\mathbf{G}_\Lambda$ and $\mathbf{G}_\Sigma$ can be computed (Adler and Horwitz, 1996) by combining Eq. (2.11c) with the definition of the bracket in Eq. (1.11a), with the result

$$\begin{aligned} \{\mathbf{G}_\Lambda, \mathbf{G}_\Sigma\} &= \text{Tr}\sum_r \epsilon_r \left(\frac{\delta \mathbf{G}_\Lambda}{\delta q_r} \frac{\delta \mathbf{G}_\Sigma}{\delta p_r} - (\Lambda \leftrightarrow \Sigma) \right) \\ &= \text{Tr}\sum_r \epsilon_r \left(-[\Lambda, p_r]\epsilon_r[\Sigma, q_r] - (\Lambda \leftrightarrow \Sigma) \right) \\ &= \text{Tr}\sum_r p_r[[\Lambda, \Sigma], q_r] = \mathbf{G}_{[\Lambda,\Sigma]}. \end{aligned} \tag{2.12a}$$

Hence the Lie algebra of the generators $\mathbf{G}_\Lambda$ under the generalized Poisson bracket is isomorphic to the algebra of the matrices Λ under commutation. Equation (2.12a), which is an analog of the "current algebra" group properties of integrated charges in quantum field theory, can be generalized (Adler and Horwitz, 1996) to an analog of the local current algebra of quantum field theory as follows. Let us write

$$\tilde{C} = \sum_r \tilde{C}_r,$$

$$\tilde{C}_r \equiv \epsilon_r q_r p_r - p_r q_r, \tag{2.12b}$$

and let us define a "local" trace generator $\mathbf{G}_{\Lambda;r}$ by

$$\mathbf{G}_{\Lambda;r} = \mathrm{Tr}\Lambda\tilde{C}_r. \tag{2.12c}$$

Then a straightforward calculation, similar to that leading to Eq. (2.12a), shows that

$$\{\mathbf{G}_{\Lambda;r}, \mathbf{G}_{\Sigma;s}\} = \delta_{rs}\mathbf{G}_{[\Lambda,\Sigma];r}. \tag{2.12d}$$

In addition to the canonical generators for global unitary transformations given in Eqs. (2.11a–c), we can also define general canonical transformations. Letting $\mathbf{G} = \mathrm{Tr}G$, with G self-adjoint but otherwise arbitrary, a general infinitesimal canonical transformation is defined, in analogy with Eq. (2.11c), by

$$\delta p_r = -\frac{\delta\mathbf{G}}{\delta q_r},$$

$$\delta q_r = \epsilon_r \frac{\delta\mathbf{G}}{\delta p_r}, \tag{2.13a}$$

which is the natural extension to trace dynamics of an infinitesimal canonical transformation in classical mechanics. In terms of the symplectic variables x_r introduced in Section 1.4, Eq. (2.13a) can be written in the compact form

$$\delta x_r = \sum_s \omega_{rs} \frac{\delta\mathbf{G}}{\delta x_s}. \tag{2.13b}$$

Letting $\mathbf{A} \equiv \mathbf{A}[\{x_r\}]$ be an arbitrary trace functional, we find immediately that to first order under a canonical transformation

$$\mathbf{A} + \delta\mathbf{A} \equiv \mathbf{A}[\{x_r + \delta x_r\}]$$

$$= \mathbf{A} + \mathrm{Tr}\sum_r \frac{\delta\mathbf{A}}{\delta x_r}\delta x_r$$

$$= \mathbf{A} + \mathrm{Tr}\sum_{r,s} \frac{\delta\mathbf{A}}{\delta x_r}\omega_{rs}\frac{\delta\mathbf{G}}{\delta x_s}$$

$$= \mathbf{A} + \{\mathbf{A}, \mathbf{G}\}, \tag{2.14a}$$

that is

$$\delta\mathbf{A} = \{\mathbf{A}, \mathbf{G}\}. \tag{2.14b}$$

Comparing Eq. (2.13b) with Eq. (1.15b), we see that when $\mathbf{G}$ is taken as $\mathbf{H}dt$, with $\mathbf{H}$ the trace Hamiltonian and dt an infinitesimal time step, then $\delta x_r = \dot{x}_r dt$ gives the small change in x_r resulting from the dynamics of the system over that time step. So, as expected, the Hamiltonian dynamics of the system is a special case of a canonical transformation.

Let us now consider canonical transformations with generators $\mathbf{G}$ which are global unitary invariant, that is, which are constructed from the $\{x_r\}$ using only c-number non-dynamical coefficients. Global unitary invariance implies that these generators obey

$$\{\mathbf{G}, \mathbf{G}_\Lambda\} = 0, \tag{2.15}$$

with $\mathbf{G}_\Lambda$ the global unitary generator of Eq. (2.11a). But using Eq. (2.14b), Eq. (2.15) has the alternate interpretation that $\mathbf{G}_\Lambda$ is invariant under a canonical transformation $\mathbf{G}$ which is global unitary invariant, and since the anti-self-adjoint matrix Λ is arbitrary, this implies that $\tilde{C}$ is invariant under any canonical transformation with a global unitary invariant generator. (This could also have been deduced by a calculation in direct analogy with Eqs. (2.7a) through (2.9b).) This invariance group of $\tilde{C}$ has the following significance. Consider a Poincaré invariant trace dynamics field theory with a global unitary invariant trace Lagrangian, examples of which are given in Sections 2.3 and 2.4 and Chapter 3 below. In such a theory, there is a set of trace functional Poincaré generators, that represent the Poincaré group under the generalized Poisson bracket operation, without any assumption of canonical commutators/anticommutators for the underlying dynamical variables $\{x_r\}$. These Poincaré generators are global unitary invariant (that is, if the Lagrangian involves only c-number non-dynamical coefficients, this property carries over to the trace energy-momentum tensor and to the trace Poincaré generators), and so we can conclude from the above discussion of canonical invariance that $\tilde{C}$ is Poincaré invariant. This will be seen explicitly in the examples given in Sections 2.3 and 2.4 and Chapter 3, and will play a role in our later analysis of the emergence of quantum behavior from the statistical dynamics of global unitary invariant matrix models.

For each phase space variable q_r, p_r, let us define the classical part q_r^c, p_r^c and the non-commutative remainder q_r', p_r', by

$$q_r^c = \frac{1}{N}\mathrm{Tr}q_r, \quad p_r^c = \frac{1}{N}\mathrm{Tr}p_r,$$
$$q_r' = q_r - q_r^c, \quad p_r' = p_r - p_r^c, \tag{2.16a}$$

so that bosonic q_r^c, p_r^c are c-numbers, fermionic q_r^c, p_r^c are Grassmann c-numbers, and the remainders are traceless

$$\mathrm{Tr}\, q_r' = \mathrm{Tr}\, p_r' = 0. \tag{2.16b}$$

Then since q_r^c, p_r^c commute (anticommute) with q_s', p_s' for r, s both bosonic (fermionic), we see that the classical parts of the phase space variables make no contribution to $\tilde{C}$, and Eq. (2.6) can be rewritten as

$$\tilde{C} = \sum_{r \in B} [q_r', p_r'] - \sum_{r \in F} \{q_r', p_r'\}. \tag{2.16c}$$

Thus $\tilde{C}$ depends only on the non-commutative parts of the matrix phase space variables.

We conclude this section by showing that the argument for the adjointness properties of $\tilde{C}$ can be generalized, allowing for the possibility that $\tilde{C}$ can have a component which is self-adjoint, when we allow a more general assignment of fermionic adjointness properties than that of Eq. (2.4b). Let us consider a trace Lagrangian which has a fermionic kinetic term of the form

$$\mathbf{L}_{\mathrm{kin}} = \mathrm{Tr} \sum_{r,s \in F} q_r^\dagger A_{rs} \dot{q}_s, \tag{2.17a}$$

with A_{rs} for each r, s an $N \times N$ constant matrix. This trace Lagrangian will be real, up to a total time derivative, provided that the set of matrices A_{rs} obeys

$$A_{rs} = A_{sr}^\dagger. \tag{2.17b}$$

The momentum p_s canonically conjugate to q_s is

$$p_s = \frac{\delta \mathbf{L}}{\delta \dot{q}_s} = \sum_{r \in F} q_r^\dagger A_{rs}, \tag{2.17c}$$

and the kinetic Lagrangian takes the form

$$\mathbf{L}_{\mathrm{kin}} = \mathrm{Tr} \sum_{s \in F} p_s \dot{q}_s, \tag{2.17d}$$

which is clearly global unitary invariant as a function of the phase space variables x_r, even though $\mathbf{L}$ was not global unitary invariant when expressed in terms of the original variables q_r, $q_r^\dagger$. Let us now suppose that the remaining terms in $\mathbf{L}$ also have the property that they are global unitary invariant when expressed in terms of the phase space variables x_r; then the trace Hamiltonian $\mathbf{H}$ will also be global unitary invariant. The argument of Eqs. (2.7b) through (2.9b), which does not make use of the adjointness assignment of Eq. (2.4b), then implies that $\tilde{C}$ of Eq. (2.6) is still conserved.

To see when the possibility of a self-adjoint component of $\tilde{C}$ can be realized, we consider the fermionic part $\tilde{C}_F$, for which we have

$$\tilde{C}_F = -\sum_{s \in F}\{q_s, p_s\} = -\sum_{r,s \in F}(q_s q_r^\dagger A_{rs} + q_r^\dagger A_{rs} q_s),$$

$$\tilde{C}_F^\dagger = \sum_{r,s \in F}(A_{rs}^\dagger q_r q_s^\dagger + q_s^\dagger A_{rs}^\dagger q_r) = \sum_{r,s \in F}(A_{rs} q_s q_r^\dagger + q_r^\dagger A_{rs} q_s), \quad (2.18a)$$

where in rewriting $\tilde{C}_F^\dagger$ in the second line we have used the condition of Eq. (2.17b). Adding the two equations, we get

$$\tilde{C}_F + \tilde{C}_F^\dagger = -\sum_{r,s \in F}[q_s q_r^\dagger, A_{rs}], \quad (2.18b)$$

showing that when the right-hand side of Eq. (2.18b) is non-zero, the operator $\tilde{C}$ is no longer anti-self-adjoint. As a direct check that the commutator on the right-hand side of Eq. (2.18b) is self-adjoint, we have

$$\left(\sum_{r,s \in F}[q_s q_r^\dagger, A_{rs}]\right)^\dagger = -\sum_{r,s \in F}[A_{rs}^\dagger, q_r q_s^\dagger] = -\sum_{r,s \in F}[A_{sr}, q_r q_s^\dagger]$$

$$= \sum_{r,s \in F}[q_r q_s^\dagger, A_{sr}] = \sum_{r,s \in F}[q_s q_r^\dagger, A_{rs}]. \quad (2.18c)$$

When A_{rs} is a c-number for all r, s, then the right-hand side of Eq. (2.18b) vanishes, and $\tilde{C}$ is anti-self-adjoint. Thus, with the c-number choice

$$A_{rs} = \delta_{rs}, \quad (2.18d)$$

which trivially satisfies the condition of Eq. (2.17b) and corresponds to the fermion kinetic structure and adjointness assignment used in Eqs. (2.4a,b), we recover our earlier conclusion that $\tilde{C}$ is anti-self-adjoint. Throughout most of this book we shall use this simple choice of A_{rs}, and we shall see that an anti-self-adjoint $\tilde{C}$ naturally leads to an emergent quantum dynamics. However, in Chapter 6, where we consider stochastic corrections to the Schrödinger equation, we shall consider the possibility that $\tilde{C}$ can have a self-adjoint part as well.

As a very simple example of a nontrivial trace action that has a conserved $\tilde{C}$ that is not anti-self-adjoint, consider

$$\mathbf{L} = \text{Tr}[q^\dagger A(\dot{q} + qB) + \frac{1}{2}\dot{B}^2], \quad (2.19a)$$

with $A = A^\dagger$ a fixed matrix and with $B = -B^\dagger$ an anti-self-adjoint bosonic operator. Then with $q_F = q$, $p_F = q^\dagger A$, and $q_B = B$, $p_B = \dot{B}$, the corresponding trace Hamiltonian is

$$\mathbf{H} = \text{Tr}(-p_F q_F q_B + \frac{1}{2}p_B^2), \quad (2.19b)$$

which is a global unitary invariant function of its arguments. Thus the operator

$$\tilde{C} = [q_B, p_B] - \{q_F, p_F\} \tag{2.20a}$$

is conserved, as can be checked explicitly by use of the operator equations of motion

$$\dot{q}_F = -q_F q_B, \quad \dot{p}_F = q_B p_F,$$

$$\dot{q}_B = p_B, \quad \dot{p}_B = p_F q_F. \tag{2.20b}$$

However, when A is not a c-number, the calculations of Eqs. (2.18b,c) show that $\tilde{C}$ has a piece that is self-adjoint

$$\tilde{C} + \tilde{C}^\dagger = [A, qq^\dagger] = [A, q_F p_F A^{-1}]. \tag{2.21a}$$

The anti-self-adjoint and self-adjoint parts of $\tilde{C}$ are separately conserved, as we readily verify from the equations of motion of Eq. (2.21)

$$\frac{d}{dt}(q_F p_F) = \dot{q}_F p_F + q_F \dot{p}_F = 0. \tag{2.21b}$$

We complete this discussion by exploring whether there is a connection between time-reversal noninvariance and the appearance of a self-adjoint piece in $\tilde{C}$. Let us define the anti-unitary time-reversal transformation $\mathcal{T}$, in analogy with the standard definition for fermion fields, by

$$\mathcal{T} i \mathcal{T}^\dagger = -i,$$

$$\mathcal{T} q_r(t) \mathcal{T}^\dagger = \sum_{s \in F} U_{rs} q_s(-t),$$

$$\mathcal{T} q_r^\dagger(t) \mathcal{T}^\dagger = \sum_{s \in F} q_s^\dagger(-t) U_{rs}^\dagger. \tag{2.22a}$$

Here U_{rs} is a unitary matrix for each r, s, and we note that under a linear superposition of the q_r with complex coefficients, one obtains an antilinear superposition of the corresponding $q_s(-t)$, that is, a superposition with complex conjugated coefficients. With this definition, we find that the transformation of the expression appearing in Eq. (2.17a) for the fermionic kinetic energy is

$$\mathcal{T}\left(\sum_{r,s \in F} q_r^\dagger(t) A_{rs} \partial_t q_s(t)\right) \mathcal{T}^\dagger$$

$$= \mathcal{T}\left(\sum_{m,n \in F} q_m^\dagger(t) A_{mn} \partial_t q_n(t)\right) \mathcal{T}^\dagger$$

$$= \sum_{r,s \in F} \sum_{m,n \in F} q_r^\dagger(-t) U_{mr}^\dagger A_{mn}^* U_{ns} \partial_t q_s(-t). \tag{2.22b}$$

Hence the kinetic action $\int dt \mathbf{L}_{\text{kin}}$ will be form-invariant under time-reversal provided

$$A_{rs} = - \sum_{m,n \in F} U_{mr}^\dagger A_{mn}^* U_{ns} = \sum_{m,n \in F} U_{mr}^\dagger A_{nm}^T U_{ns}, \qquad (2.22c)$$

where in the second equality we have substituted the condition of Eq. (2.17b), and where the superscript T denotes a transpose of the matrix structure of A_{nm}, but does not act on the summation indices m, n. Referring to Eq. (2.18a) and applying the transformations of Eq. (2.22a) to evaluate $T\tilde{C}_F(t)T^\dagger$, we find that

$$T\tilde{C}_F(t)T^\dagger + \tilde{C}_F(-t) = \sum_{r,s \in F} \Big(-q_s(-t)q_r^\dagger(-t)A_{rs}$$

$$- \sum_{m,n \in F} U_{ns} q_s(-t) q_r^\dagger(-t) U_{mr}^\dagger A_{mn}^* \Big). \qquad (2.22d)$$

When Eq. (2.22c) can be satisfied by matrices U_{rs} that are c-numbers for each r, s, the factor U_{ns} in the second term on the right-hand side of Eq. (2.22d) can be commuted through to the right. We can then use the condition of Eq. (2.22c) to conclude that the two terms on the right-hand side of Eq. (2.22d) cancel, so that $T\tilde{C}_F(t)T^\dagger + \tilde{C}_F(-t) = 0$ and $\tilde{C}_F$ is time-reversal odd. Since this is also the behavior expected for the bosonic part of $\tilde{C}$ (bosonic qs are time-reversal even, while bosonic ps are time-reversal odd), we conclude that $\tilde{C}$ is time-reversal odd when Eq. (2.22c) can be satisfied with c-number matrices U_{rs}.

The condition of Eq. (2.22c) is evidently satisfied by the c-number matrix $U_{rs} = \delta_{rs}$ when we make the simplest choice $A_{rs} = \delta_{rs}$ of Eq. (2.18d). So in this case, which corresponds to the simplest fermionic adjointness assignment of Eq. (2.4b), $\tilde{C}$ is purely anti-self-adjoint and is time-reversal odd. If we instead take the somewhat more general choice $A_{rs} = A\delta_{rs}$, with $A = -A^\dagger$, and also take U_{rs} to have the form $U_{rs} = \delta_{rs}U$, then Eq. (2.22c) simplifies to

$$A = -U^\dagger A^* U = -U^\dagger A^T U, \qquad (2.22e)$$

with the superscript T denoting the matrix transpose. This condition can be satisfied, but requires U to have a nontrivial matrix structure. For example, if we let $\sigma_{1,2,3}$ denote the standard Pauli spin matrices, and take $A = A^\dagger = A^T = \sigma_3$, then the condition of Eq. (2.22e) is satisfied by taking either $U = \sigma_1$ or $U = \sigma_2$. In this case U_{ns} cannot be commuted through to the right in the second term on the right-hand side of Eq. (2.22d), and $\tilde{C}$ has a term which is time-reversal even. We conclude that the generalization of the fermionic kinetic term and adjointness assignment that leads to the presence of a self-adjoint piece in $\tilde{C}$ also, in general, leads to time-reversal violation.

To summarize the analysis so far, associated with a generic trace dynamics model there are three generic conserved quantities, irrespective of whether the degree of freedom index r plays the role of a spatial index. The trace Hamiltonian **H** is always conserved, independent of structural assumptions. When each mono-mial in **H** or **L** has equal numbers of fermionic operators p_r and q_s, there is a conserved trace "fermion number" **N** given by Eq. (2.2a). When **H** is global unitary invariant, the corresponding Noether charge gives a conserved operator $\tilde{C}$ with the dimensions of action, given by Eq. (2.6). When fermionic adjointness is assigned in the simplest manner, as in Eq. (2.4b), the operator $\tilde{C}$ is anti-self-adjoint and is odd under time-reversal. For more general fermionic adjointness assignments, as in Eqs. (2.17a–d), the operator $\tilde{C}$ can have a self-adjoint part, and there is in general a violation of time-reversal symmetry.

2.3 Conserved quantities for continuum spacetime theories

Up to this point we have taken the index r labeling degrees of freedom to be a discrete index. Let us turn now to the case of continuum spacetime theories, in which r is a composite label indicating the spatial point $\vec{x}$ and the field variable q_ℓ evaluated at that point, so that we have the correspondence $q_r(t) \leftrightarrow q_\ell(\vec{x}, t) \equiv q_\ell(x)$, where in the final expression we have adopted a four-vector notation $(\vec{x}, t = x^0) = x$. We shall consider in this section spacetime theories which are trace dynamics analogs of standard local field theories, in which the trace Lagrangian is constructed only from the field variables $q_\ell(x)$ and their first spacetime derivatives $\partial_\mu q_\ell(x) = \partial q_\ell(x)/\partial x^\mu$. Thus we have

$$\mathbf{L} = \int d^3x \mathcal{L}(\{q_\ell(x)\}, \{\partial_\mu q_\ell(x)\}), \tag{2.23a}$$

with $\mathcal{L}$ the trace Lagrangian density. Requiring the trace action defined by Eq. (1.6b) to be invariant with respect to variations of the fields that vanish at spatial infinity, we have

$$0 = \delta \mathbf{S} = \int d^4x \mathrm{Tr} \sum_\ell \left(\frac{\delta \mathcal{L}}{\delta q_\ell(x)} \delta q_\ell(x) + \frac{\delta \mathcal{L}}{\delta \partial_\mu q_\ell(x)} \delta \partial_\mu q_\ell(x) \right), \tag{2.23b}$$

which after an integration by parts gives

$$0 = \int d^4x \mathrm{Tr} \sum_\ell \left(\frac{\delta \mathcal{L}}{\delta q_\ell(x)} - \partial_\mu \frac{\delta \mathcal{L}}{\delta \partial_\mu q_\ell(x)} \right) \delta q_\ell(x). \tag{2.23c}$$

Requiring this to hold for general same-type operator variations $\delta q_\ell(x)$ then gives the operator Euler–Lagrange equations

$$\frac{\delta \mathcal{L}}{\delta q_\ell(x)} - \partial_\mu \frac{\delta \mathcal{L}}{\delta \partial_\mu q_\ell(x)} = 0. \tag{2.23d}$$

Working from the trace Lagrangian of Eq. (2.23a), let us now derive analogs of the standard Noether's theorem, by considering transformations that leave the trace Lagrangian invariant. We begin by considering an internal symmetry transformation with an infinitesimal c-number parameter $\alpha(x)$, under which the dynamical variables transform as

$$q_\ell(x) \to q_\ell(x) + \alpha(x)\Delta_\ell(x),$$
$$\partial_\mu q_\ell(x) \to \partial_\mu q_\ell(x) + \alpha(x)\partial_\mu\Delta_\ell(x) + \partial_\mu\alpha(x)\Delta_\ell(x), \tag{2.24a}$$

with $\Delta_\ell(x)$ specified functions constructed from the dynamical variables. Substituting Eq. (2.24a) into the trace Lagrangian, we see that the change in the trace Lagrangian is given by

$$\mathbf{L} \to \mathbf{L} + \int d^3x \sum_\ell \mathrm{Tr}\left[\alpha(x)\left(\frac{\delta\mathcal{L}}{\delta q_\ell(x)}\Delta_\ell(x) + \frac{\delta\mathcal{L}}{\delta\partial_\mu q_\ell(x)}\partial_\mu\Delta_\ell(x) \right) \right.$$
$$\left. + \partial_\mu\alpha(x)\frac{\delta\mathcal{L}}{\delta\partial_\mu q_\ell(x)}\Delta_\ell(x) \right]. \tag{2.24b}$$

From the coefficient of $\partial_\mu\alpha(x)$, we extract a local trace current $\mathbf{J}^\mu$ given by

$$\mathbf{J}^\mu(x) = \mathrm{Tr}\sum_\ell \frac{\delta\mathcal{L}}{\delta\partial_\mu q_\ell(x)}\Delta_\ell(x). \tag{2.24c}$$

Calculating the divergence of this current and simplifying by use of the Euler–Lagrange equations of Eq. (2.23c), we get

$$\partial_\mu\mathbf{J}^\mu(x) = \mathrm{Tr}\sum_\ell \left(\partial_\mu\frac{\delta\mathcal{L}}{\delta\partial_\mu q_\ell(x)}\Delta_\ell(x) + \frac{\delta\mathcal{L}}{\delta\partial_\mu q_\ell(x)}\partial_\mu\Delta_\ell(x) \right)$$
$$= \mathrm{Tr}\sum_\ell \left(\frac{\delta\mathcal{L}}{\delta q_\ell(x)}\Delta_\ell(x) + \frac{\delta\mathcal{L}}{\delta\partial_\mu q_\ell(x)}\partial_\mu\Delta_\ell(x) \right). \tag{2.24d}$$

We see that the right-hand side of Eq. (2.24d) is identical with the coefficient of $\alpha(x)$ in the change in the trace Lagrangian in Eq. (2.24b). Hence when the trace Lagrangian is invariant under the transformation of Eq. (2.24a) for constant $\alpha(x)$, the trace current $\mathbf{J}^\mu$ defined by the variation of the trace Lagrangian for non-constant $\alpha(x)$ is conserved.

As an application, let us consider a trace Lagrangian with a fermionic structure analogous to that of Eq. (2.5c)

$$\mathbf{L} = \mathrm{Tr}\int d^3x \sum_{\ell\in F} \left(-\overline{\psi}_\ell(x)\gamma^\mu\partial_\mu\psi_\ell(x) \right)$$
$$+ \text{ terms with no time derivatives of fermions} \tag{2.25a}$$

where γ^μ are the Dirac gamma matrices, $\overline{\psi}_\ell = \psi_\ell^\dagger\gamma^0$, and where in our metric conventions $(\gamma^0)^2 = -1$. Suppose that this trace Lagrangian is invariant under the fermionic rephasings $\psi_\ell \to \exp(i\alpha)\psi_\ell$ and $\psi^\dagger_{\ell'} \to \exp(-i\alpha)\psi^\dagger_{\ell'}$, for arbitrary ℓ, ℓ' and with constant α. This will be true if in each term of $\mathbf{L}$, for each factor ψ_ℓ there is also a factor $\psi^\dagger_{\ell'}$. When α is not a constant, to first order the trace Lagrangian will then change under this rephasing according to

$$\mathbf{L} \to \mathbf{L} - i \int d^3x \partial_\mu \alpha \mathrm{Tr} \sum_{\ell \in F} \overline{\psi}_\ell(x) \gamma^\mu \psi_\ell(x). \tag{2.25b}$$

The Noether analysis of Eqs. (2.24a–d) then tells us that the coefficient of $-\partial_\mu \alpha$ in Eq. (2.25b) gives a conserved trace current $\mathbf{N}^\mu(x)$

$$\mathbf{N}^\mu(x) = i\mathrm{Tr} \sum_{\ell \in F} \overline{\psi}_\ell(x) \gamma^\mu \psi_\ell(x), \tag{2.25c}$$

which obeys

$$\partial_\mu \mathbf{N}^\mu(x) = 0, \tag{2.25d}$$

so that the spatial integral of the time or zero component of this current gives a conserved trace charge $\mathbf{N}$

$$\mathbf{N} = \int d^3x \mathbf{N}^0(x) = -i \int d^3x \sum_{\ell \in F} \psi^\dagger_\ell(x)\psi_\ell(x), \quad \dot{\mathbf{N}} = \partial_0 \mathbf{N} = 0. \tag{2.25e}$$

Let us next consider the Noether's theorem analog that follows when the trace Lagrangian is global unitary invariant. In this case the relevant infinitesimal transformation has an operator-valued parameter $\Lambda(x)$, and from Eq. (2.8a) takes the form

$$q_\ell(x) \to q_\ell(x) - [\Lambda(x), q_\ell(x)],$$
$$\partial_\mu q_\ell(x) \to \partial_\mu q_\ell(x) - [\Lambda(x), \partial_\mu q_\ell(x)] - [\partial_\mu \Lambda(x), q_\ell(x)]. \tag{2.26a}$$

Substituting the transformation of Eq. (2.26a) into the trace Lagrangian, and using the cyclic identities to permute factors of Λ and $\partial_\mu \Lambda$ to the right, we find that the trace Lagrangian transforms according to

$$\mathbf{L} \to \mathbf{L} - \mathrm{Tr} \int d^3x (\tilde{D}\Lambda + \tilde{C}^\mu \partial_\mu \Lambda), \tag{2.26b}$$

where we have defined

$$\tilde{D} = \sum_\ell \left(\epsilon_\ell q_\ell \frac{\delta\mathcal{L}}{\delta q_\ell} - \frac{\delta\mathcal{L}}{\delta q_\ell} q_\ell + \epsilon_\ell \partial_\mu q_\ell \frac{\delta\mathcal{L}}{\delta \partial_\mu q_\ell} - \frac{\delta\mathcal{L}}{\delta \partial_\mu q_\ell} \partial_\mu q_\ell \right),$$

$$\tilde{C}^\mu = \sum_\ell \left(\epsilon_\ell q_\ell \frac{\delta\mathcal{L}}{\delta \partial_\mu q_\ell} - \frac{\delta\mathcal{L}}{\delta \partial_\mu q_\ell} q_\ell \right). \tag{2.26c}$$

Calculating the divergence of $\tilde{C}^\mu$, and simplifying by using the Euler–Lagrange equations of Eq. (2.23d) then gives

$$\partial_\mu \tilde{C}^\mu = \tilde{D}. \tag{2.26d}$$

Thus, when the trace Lagrangian is global unitary invariant for constant Λ, which requires the vanishing of $\tilde{D}$, the current $\tilde{C}^\mu$, obtained as the coefficient of $-\partial_\mu \Lambda(x)$ in the variation of **L** for non-constant $\Lambda(x)$, is conserved. The corresponding conserved charge is

$$\tilde{C} = \int d^3x \, \tilde{C}^0 = \int d^3x \sum_\ell (\epsilon_\ell q_\ell p_\ell - p_\ell q_\ell), \tag{2.27a}$$

where in the continuum case the canonical momentum is defined by

$$p_\ell(x) = \frac{\delta \mathcal{L}}{\delta \partial_0 q_\ell(x)}. \tag{2.27b}$$

When the trace Lagrangian density $\mathcal{L}$ is Poincaré invariant, there are further generic conservation laws, beyond the ones just discussed that are associated with internal symmetry invariances. We derive these by following the standard textbook treatments (see, e.g., Bjorken and Drell, 1965 and Weinberg, 1995). We consider first the consequences of a translation of the system by an infinitesimal constant displacement a^μ, so that

$$\mathcal{L}(x) = \mathcal{L}(\{q_\ell(x)\}, \{\partial_\mu q_\ell(x)\}) \tag{2.28a}$$

is shifted to

$$\begin{aligned} \mathcal{L}(x+a) &= \mathcal{L}(\{q_\ell(x+a)\}, \{\partial_\mu q_\ell(x+a)\}) \\ &\simeq \mathcal{L}(\{q_\ell + a^\sigma \partial_\sigma q_\ell\}, \{\partial_\mu q_\ell + a^\sigma \partial_\sigma \partial_\mu q_\ell\}). \end{aligned} \tag{2.28b}$$

Subtracting Eq. (2.28a) from Eq. (2.28b) and using Eq. (2.23d), we get to first order in a^μ the identity

$$\begin{aligned} a^\sigma \partial_\sigma \mathcal{L}(x) &= \mathrm{Tr} \sum_\ell \left(\frac{\delta \mathcal{L}}{\delta q_\ell} a^\sigma \partial_\sigma q_\ell + \frac{\delta \mathcal{L}}{\delta \partial_\mu q_\ell} a^\sigma \partial_\sigma \partial_\mu q_\ell \right) \\ &= \mathrm{Tr} \sum_\ell \left((\partial_\mu \frac{\delta \mathcal{L}}{\delta \partial_\mu q_\ell}) a^\sigma \partial_\sigma q_\ell + \frac{\delta \mathcal{L}}{\delta \partial_\mu q_\ell} a^\sigma \partial_\sigma \partial_\mu q_\ell \right) \\ &= a^\sigma \partial_\mu \mathrm{Tr} \sum_\ell \left(\frac{\delta \mathcal{L}}{\delta \partial_\mu q_\ell} \partial_\sigma q_\ell \right). \end{aligned} \tag{2.28c}$$

Since a^σ is arbitrary, factoring it away gives the identity

$$\partial_\mu T^{\mu\sigma} = 0, \tag{2.29a}$$

with the trace energy-momentum tensor density $T^{\mu\sigma}$ given by

$$T^{\mu\sigma} = \eta^{\mu\sigma}\mathcal{L} - \text{Tr}\sum_{\ell}\frac{\delta\mathcal{L}}{\delta\partial_\mu q_\ell}\partial^\sigma q_\ell, \qquad (2.29b)$$

where $\eta^{\mu\sigma} = \text{diag}(1, 1, 1, -1)$ is the Minkowski metric. Conservation of the density $T^{\mu\sigma}$ implies that there is a conserved trace energy-momentum four-vector $\mathbf{P}^\sigma$ defined by

$$\mathbf{P}^\sigma = \int d^3x\, T^{0\sigma}, \qquad (2.29c)$$

the time component of which gives the conserved trace Hamiltonian

$$\mathbf{H} = \mathbf{P}^0 = -\mathbf{L} + \int d^3x\, \text{Tr}\frac{\delta\mathcal{L}}{\delta\partial_0 q_\ell}\partial_0 q_\ell. \qquad (2.30)$$

We consider next the consequences of a four-space rotation of the system, with an infinitesimal rotation parameter $\omega_{\mu\nu} = -\omega_{\nu\mu}$, so that the coordinate x_ν is shifted to $x'_\nu = x_\nu + \omega_{\nu\mu}x^\mu$. Under this shift, the field $q_\ell(x)$ is shifted by an amount

$$\delta q_\ell = q_\ell(x') - q_\ell(x) - (1/2)\omega_{\mu\nu}\sum_m \Sigma^{\mu\nu}_{\ell m}q_m(x), \qquad (2.31a)$$

with $\Sigma_{\ell m}$ a matrix characterizing the intrinsic spin structure associated with the field q_ℓ. By reasoning identical to that used in the case of a translational shift, we find the formula

$$\delta\mathcal{L} = \omega_{\sigma\mu}x^\mu\partial^\sigma\mathcal{L}$$
$$= \text{Tr}\sum_{\ell}\partial_\lambda\left(\frac{\delta\mathcal{L}}{\delta\partial_\lambda q_\ell}\delta q_\ell\right). \qquad (2.31b)$$

Substituting Eq. (2.31a) for δq_ℓ and doing some algebra, and using the fact that the rotation parameter $\omega_{\sigma\mu}$ is a general antisymmetric tensor, we get the identity

$$\partial_\lambda\mathcal{M}^{\lambda\sigma\mu} = 0, \qquad (2.32a)$$

with the trace generalized angular momentum density $\mathcal{M}^{\lambda\sigma\mu}$ given by

$$\mathcal{M}^{\lambda\sigma\mu} = (x^\mu\eta^{\sigma\lambda} - x^\sigma\eta^{\mu\lambda})\mathcal{L} + \text{Tr}\sum_{\ell}\frac{\delta\mathcal{L}}{\delta\partial_\lambda q_\ell}\left[(x^\sigma\partial^\mu - x^\mu\partial^\sigma)q_\ell + \sum_m \Sigma^{\sigma\mu}_{\ell m}q_m\right]$$

$$= x^\mu T^{\lambda\sigma} - x^\sigma T^{\lambda\mu} + \text{Tr}\sum_{\ell m}\frac{\delta\mathcal{L}}{\delta\partial_\lambda q_\ell}\Sigma^{\sigma\mu}_{\ell m}q_m. \qquad (2.32b)$$

From the conservation of the trace density $\mathcal{M}^{\lambda\sigma\mu}$, we learn that the spatial integral of the $\lambda = 0$ component of this density is a conserved charge

$$\mathbf{M}^{\sigma\mu} = \int d^3x \, \mathcal{M}^{0\sigma\mu}, \tag{2.32c}$$

which together with $\mathbf{P}^\sigma$ gives the complete set of Poincaré generators. Since we have seen that the generalized Poisson bracket of any two conserved charges is also a conserved charge, and since there are no further bosonic charges associated with Poincaré invariance, we know on general grounds that the trace Poincaré generators will form a closed Lie algebra under the action of the generalized Poisson bracket. Clearly, this Lie algebra will be in fact the Lie algebra of the Poincaré group, since it must be isomorphic to the algebra of the translational and rotational transformations parameterized by a^μ and $\omega^{\mu\nu}$, but verifying this by direct computation of generalized Poisson brackets in any specific model involves, in general, a great deal of algebra.

The canonical formalism for constructing $\mathcal{T}^{\mu\nu}$ gives an energy-momentum tensor which is not symmetric in its indices, whereas the trace energy-momentum tensor that couples to gravitation through the metric must be symmetric. This problem can be dealt with, as in the standard field theory context, by adding a suitable conserved asymmetric trace tensor to $\mathcal{T}^{\mu\nu}$, so as to give a sum that is symmetric, conserved, and leads to no change in the trace four-momentum $\mathbf{P}^\nu$. Let $\Theta^{[\mu\rho]\nu}(x)$ be a trace tensor that is antisymmetric in the indices μ and ρ, and let us form the total energy-momentum tensor $\mathcal{T}_{\text{tot}}^{\mu\nu}$ according to

$$\mathcal{T}_{\text{tot}}^{\mu\nu} = \mathcal{T}^{\mu\nu} + \partial_\rho \Theta^{[\mu\rho]\nu}. \tag{2.33a}$$

Because of the antisymmetry of Θ in μ and ρ, the total tensor thus defined is still conserved on the μ index

$$\partial_\mu \mathcal{T}_{\text{tot}}^{\mu\nu} = \partial_\mu \mathcal{T}^{\mu\nu} + \partial_\mu \partial_\rho \Theta^{[\mu\rho]\nu} = 0. \tag{2.33b}$$

Also, because of the antisymmetry of Θ, when $\mu = 0$ the added term contains only spatial derivative terms, which vanish when integrated over three space, and so the energy-momentum four-vector calculated from $\mathcal{T}_{\text{tot}}^{\mu\nu}$ is the same as that calculated from $\mathcal{T}^{\mu\nu}$.

Let us now show that we can construct a Θ that renders $\mathcal{T}_{\text{tot}}^{\mu\nu}$ symmetric. The basic observation is that Eqs. (2.32a,b) and (2.29a), when combined, give the following expression for the antisymmetric part of $\mathcal{T}^{\mu\nu}$

$$-\frac{1}{2}(\mathcal{T}^{\mu\nu} - \mathcal{T}^{\nu\mu}) = \frac{1}{2}\partial_\rho \text{Tr} \sum_{\ell m} \frac{\delta\mathcal{L}}{\delta\partial_\rho q_\ell} \Sigma_{\ell m}^{\nu\mu} q_m. \tag{2.34a}$$

This motivates the choice

$$\Theta^{[\mu\rho]\nu} = \frac{1}{2}\text{Tr}\sum_{\ell m}\left(\frac{\delta\mathcal{L}}{\delta\partial_\rho q_\ell}\Sigma^{\nu\mu}_{\ell m}q_m - \frac{\delta\mathcal{L}}{\delta\partial_\mu q_\ell}\Sigma^{\nu\rho}_{\ell m}q_m - \frac{\delta\mathcal{L}}{\delta\partial_\nu q_\ell}\Sigma^{\mu\rho}_{\ell m}q_m\right), \quad (2.34\text{b})$$

which is antisymmetric in μ and ρ, and the final two terms of which are symmetric in μ and ν. Thus by Eq. (2.34a), we have

$$\partial_\rho\Theta^{[\mu\rho]\nu} = -\frac{1}{2}(T^{\mu\nu} - T^{\nu\mu}) + \text{symmetric in }\mu\text{ and }\nu, \quad (2.34\text{c})$$

and so when $\partial_\rho\Theta^{[\mu\rho]\nu}$ is added to $T^{\mu\nu}$, it just removes the antisymmetric part of $T^{\mu\nu}$, leaving an expression that is completely symmetric in μ and ν.

There is one further "improvement" that can be made to the canonical energy-momentum tensor. If ϕ is a self-adjoint scalar field, then

$$\Delta T^{\mu\nu} \equiv -(\partial^\mu\partial^\nu - \partial_\rho\partial^\rho\eta^{\mu\nu})\text{Tr}\phi^2 \quad (2.35\text{a})$$

has the correct dimensions to be an addition to the energy-momentum tensor, and is symmetric, divergenceless on both the μ and ν indices, and when $\mu = 0$ involves only spatial derivatives. Hence any numerical multiple of Eq. (2.35a) can be freely added to the canonical energy-momentum tensor. The theory of scale and conformal invariance (Callan, Coleman, and Jackiw, 1970) shows, in fact, that when scalar fields are present, for each self-adjoint scalar field we must add Eq. (2.35a), multiplied by a factor of $1/6$, to the canonical energy-momentum tensor, in order to get the energy-momentum tensor for a scalar theory that is scale and conformal invariant in the massless limit. We shall not give the proof of this assertion, but will illustrate it with a concrete example in the next section.

2.4 An illustrative example: a Dirac fermion coupled to a scalar Klein–Gordon field

To gain familiarity with the trace dynamics formalism, and to illustrate the conserved quantities introduced in Sections 2.1 through 2.3, we formulate in this section a trace dynamics analog of a simple field theory model, in which a Dirac fermion field ψ is coupled to a self-adjoint scalar Klein–Gordon field ϕ. Since these are taken in trace dynamics to be matrix operator fields, $\phi(x)$ is a general $N \times N$ self-adjoint Grassmann even grade matrix, and each spinor component $\psi_r(x)$ of $\psi(x)$ is a general $N \times N$ Grassmann odd grade matrix. Letting Tr denote the trace over the N-dimensional matrix space, the trace Lagrangian for this model is

$$\mathbf{L} = \int d^3x\,\text{Tr}\left[-\frac{1}{2}\partial_\mu\phi\partial^\mu\phi - \frac{1}{2}m^2\phi^2 - \overline{\psi}\gamma^\mu\partial_\mu\psi + m\overline{\psi}\psi\right.$$
$$\left. -\frac{\lambda}{24}\phi^4 + g_1\overline{\psi}\phi\psi + g_2\overline{\psi}\psi\phi\right] \quad (2.36\text{a})$$

with $\overline{\psi} = \psi^\dagger \gamma^0$ as in Eq. (2.25a). In classical field theory with commutative fields the interaction terms with couplings g_1 and g_2 would not be distinct, but here, since ϕ and the spinor components of ψ are generic non-commuting matrices, the two orderings represent different interactions. Forming the variation of the trace action **S**, integrating by parts, and using the cyclic identities, we find

$$\delta \mathbf{S} = \int d^4x \, \mathrm{Tr}(A_\phi \delta\phi + \delta\overline{\psi} A_\psi + A_{\overline{\psi}} \delta\psi), \qquad (2.36b)$$

and so the Euler–Lagrange equations of motion take the form

$$0 = A_\phi = \partial_\mu \partial^\mu \phi - m^2 \phi - \frac{\lambda}{6}\phi^3 - g_1 \psi^T \overline{\psi}^T + g_2 \overline{\psi}\psi,$$
$$0 = A_\psi = -\gamma^\mu \partial_\mu \psi + m\psi + g_1 \phi \psi + g_2 \psi \phi,$$
$$0 = A_{\overline{\psi}} = \partial_\mu \overline{\psi} \gamma^\mu + m\overline{\psi} + g_1 \overline{\psi}\phi + g_2 \phi \overline{\psi}. \qquad (2.36c)$$

Here, and in the following equations, the notation ψ^T and $\overline{\psi}^T$ indicates transposition of ψ and $\overline{\psi}$ with respect to their Dirac spinor indices only, with no action on the matrix structure of the Dirac spinor components.

We can now go over to a Hamiltonian formalism, by exhibiting the time derivative terms of the trace Lagrangian

$$\mathbf{L} = \int d^3x \left[\frac{1}{2}(\dot\phi)^2 - \overline{\psi}\gamma^0 \dot\psi + \dots \right], \qquad (2.37a)$$

which shows that the canonical momenta are

$$p_\phi = \dot\phi, \quad p_\psi = -\overline{\psi}\gamma^0 = \psi^\dagger, \quad p_{\overline{\psi}} = 0. \qquad (2.37b)$$

Thus we find that the trace Hamiltonian is

$$\mathbf{H} = \int d^3x \, \mathrm{Tr}(p_\phi \dot\phi + p_\psi \dot\psi) - \mathbf{L}$$
$$= \int d^3x \, \mathrm{Tr}\left[\frac{1}{2}((\dot\phi)^2 + (\vec\nabla\phi)^2 + m^2\phi^2) + \overline{\psi}\vec\gamma \cdot \vec\partial \psi - m\overline{\psi}\psi \right.$$
$$\left. + \frac{\lambda}{24}\phi^4 - g_1\overline{\psi}\phi\psi - g_2\overline{\psi}\psi\phi \right], \qquad (2.37c)$$

and the conservation of this can be verified directly by using the Euler–Lagrange equations of Eq. (2.36c).

The trace Lagrangian of Eq. (2.36a) has equal numbers of ψ and $\overline{\psi}$ factors in each term, and is global unitary invariant, and so will have the generic conserved internal symmetry currents discussed in Section 2.3. To identify the trace fermion number current, we replace $\psi \to \psi + i\alpha\psi$ and $\overline{\psi} \to \overline{\psi} - i\alpha\overline{\psi}$ in the trace Lagrangian, and pick out the coefficient of $-\partial_\mu \alpha$. This gives

$$\mathbf{N}^\mu = i\,\mathrm{Tr}\,\overline{\psi}\gamma^\mu\psi, \qquad (2.38a)$$

which by use of the Euler–Lagrange equations is seen to be conserved

$$\partial_\mu \mathbf{N}^\mu = 0. \tag{2.38b}$$

The corresponding conserved trace fermion number is

$$\mathbf{N} = \int d^3x \mathbf{N}^0 = -i \int d^3x \mathrm{Tr}\psi^\dagger \psi. \tag{2.38c}$$

Similarly, to identify the current $\tilde{C}^\mu$ associated with global unitary invariance, we substitute $\phi \to \phi - [\Lambda, \phi]$, $\psi \to \psi - [\Lambda, \psi]$, and $\overline{\psi} \to \overline{\psi} - [\Lambda, \overline{\psi}]$ into the trace Lagrangian, and pick out the coefficient of $-\partial_\mu \Lambda$. This gives

$$\tilde{C}^\mu = -[\phi, \partial^\mu \phi] + \psi^T (\overline{\psi}\gamma^\mu)^T + \overline{\psi}\gamma^\mu \psi, \tag{2.39a}$$

which again by use of the Euler–Lagrange equations is seen to be conserved

$$\partial_\mu \tilde{C}^\mu = 0. \tag{2.39b}$$

The corresponding conserved charge is

$$\tilde{C} = \int d^3x \tilde{C}^0 = \int d^3x(-[\phi, \partial^0 \phi] + \psi^T (\overline{\psi}\gamma^0)^T + \overline{\psi}\gamma^0 \psi)$$

$$= \int d^3x \left([\phi, \dot{\phi}] - (\psi^T \psi^{\dagger T} + \psi^\dagger \psi)\right)$$

$$= \int d^3x([\phi, p_\phi] - \{\psi, p_\psi\}), \tag{2.39c}$$

in agreement with the general recipe of Eq. (2.6).

We turn next to properties of the energy-momentum tensor. Since the Dirac matrix manipulations needed to symmetrize the fermionic part of the energy-momentum tensor are not relevant for the points we wish to illustrate, let us simplify the model so that we are dealing just with the trace dynamics analog of the scalar ϕ^4 model, with trace Lagrangian density

$$\mathcal{L} = \mathrm{Tr}\left[-\frac{1}{2}\partial_\mu\phi\partial^\mu\phi - \frac{1}{2}m^2\phi^2 - \frac{\lambda}{24}\phi^4\right], \tag{2.40a}$$

and with Euler–Lagrange equations

$$0 = \partial_\mu \partial^\mu \phi - m^2 \phi - \frac{\lambda}{6}\phi^3. \tag{2.40b}$$

The symmetrized, "improved" energy-momentum tensor for this model is

$$T^{\mu\nu} = \eta^{\mu\nu}\mathcal{L} + \mathrm{Tr}\partial^\mu\phi\partial^\nu\phi - \frac{1}{6}(\partial^\mu\partial^\nu - \partial_\sigma\partial^\sigma\eta^{\mu\nu})\mathrm{Tr}\phi^2. \tag{2.40c}$$

This is readily verified, by use of the Euler–Lagrange equation of Eq. (2.40b) and the cyclic identities, to be conserved

$$\partial_\mu T^{\mu\nu} = \partial_\nu T^{\mu\nu} = 0, \tag{2.40d}$$

and to have a Lorentz trace that is given just by the scalar field mass term

$$\eta_{\mu\nu} T^{\mu\nu} = -m^2 \mathrm{Tr}\phi^2. \tag{2.40e}$$

We note, however, that there is no *operator* energy-momentum tensor that is both conserved and has a trace that vanishes when the mass m is zero. To see this, let us form the obvious symmetrized candidate

$$t^{\mu\nu} = \eta^{\mu\nu}\left[-\frac{1}{2}\partial_\sigma\phi\partial^\sigma\phi - \frac{1}{2}m^2\phi^2 - \frac{\lambda}{24}\phi^4 \right] + \frac{1}{2}(\partial^\mu\phi\partial^\nu\phi + \partial^\nu\phi\partial^\mu\phi)$$
$$-\frac{1}{6}(\partial^\mu\partial^\nu - \partial_\sigma\partial^\sigma\eta^{\mu\nu})\phi^2, \tag{2.41a}$$

so that

$$T^{\mu\nu} = \mathrm{Tr}t^{\mu\nu}. \tag{2.41b}$$

Then we find that

$$\eta_{\mu\nu}t^{\mu\nu} = -m^2\phi^2, \tag{2.41c}$$

so that the trace vanishes when $m = 0$, but the operator candidate of Eq. (2.41a) is not conserved

$$\partial_\mu t^{\mu\nu} = \frac{\lambda}{24}[(\partial^\nu\phi)\phi^3 - \phi(\partial^\nu\phi)\phi^2 - \phi^2(\partial^\nu\phi)\phi + \phi^3\partial^\nu\phi]. \tag{2.41d}$$

The right-hand side of Eq. (2.41d) is non-vanishing because in general the matrix $\partial^\nu\phi$ does not commute with the matrix ϕ, but by the cyclic identities, it vanishes inside a trace. Hence in trace dynamics, although there is always a conserved trace energy-momentum tensor, in general there is no conserved operator analog. In Chapter 7 we will relate this fact to a suggestion that the cosmological constant problem, which has not been solvable within quantum field theory, may find a solution in an underlying trace dynamics. (This suggestion was made in Adler (1997c), but the attempt there to give a calculation of the induced cosmological constant is not correct, since it did not take account of the necessity, discussed in the Introduction and in Section 4.5 below, to fix a residual global unitary invariance of the canonical ensemble.)

2.5 Symmetries of conserved quantities under $p_F \leftrightarrow q_F$

We close this chapter by discussing the symmetries of the various generic conserved quantities under the interchange, within each fermionic canonical pair, of the canonical coordinate with the corresponding canonical momentum. We begin with the trace fermion number $\mathbf{N}$, defined in Eq. (2.2a) by

$$\mathbf{N} = -i \operatorname{Tr} \sum_{r \in F} p_r q_r. \tag{2.42a}$$

If we interchange $p_r \leftrightarrow q_r$ for each fermionic canonical pair, this becomes

$$\mathbf{N} \to -i \operatorname{Tr} \sum_{r \in F} q_r p_r = i \operatorname{Tr} \sum_{r \in F} p_r q_r = -\mathbf{N}, \tag{2.42b}$$

where we have used the cyclic property of the trace given in Eq. (1.1b). Thus $\mathbf{N}$ is odd under interchange of fermionic canonical coordinates and momenta. Turning next to the generic conserved operator $\tilde{C}$, since Eq. (2.6) tells us that the fermionic contribution to $\tilde{C}$ is

$$\tilde{C}_F = -\sum_{r \in F} \{q_r, p_r\}, \tag{2.42c}$$

we learn that $\tilde{C}$ is even under the interchange of fermionic canonical coordinates and momenta.

Finally, let us consider the trace Hamiltonian $\mathbf{H}$. Here the behavior under $p_F \leftrightarrow q_F$ is structure dependent. If we consider the bilinear $\mathbf{H}_F$ introduced in Eq. (2.1a), with

$$\mathbf{H}_F = \operatorname{Tr} \sum_{r,s \in F} (p_r q_s B_{1rs} + p_r B_{2rs} q_s), \tag{2.43a}$$

then under interchange of all p_F with q_F this becomes

$$\mathbf{H}_F \to \operatorname{Tr} \sum_{r,s \in F} (q_r p_s B_{1rs} + q_r B_{2rs} p_s)$$

$$= -\operatorname{Tr} \sum_{r,s \in F} (p_s B_{1rs} q_r + p_s q_r B_{2rs}) = -\operatorname{Tr} \sum_{r,s \in F} (p_r q_s B_{2sr} + p_r B_{1sr} q_s). \tag{2.43b}$$

Hence $\mathbf{H}_F$ will be even under the interchange if we have

$$B_{1rs} = -B_{2sr}, \quad \text{all } r, s, \tag{2.43c}$$

and will be odd under the interchange if

$$B_{1rs} = B_{2sr}, \quad \text{all } r, s. \tag{2.43d}$$

The even case of Eq. (2.43c) is physically relevant for continuum field theories in which r is a spacetime box label. To see this we note that the standard Dirac fermion kinetic Hamiltonian

$$\mathbf{H}_{F \text{ kin}} = \int d^3x \, \text{Tr}\overline{\psi}\vec{\gamma} \cdot \vec{\partial}\psi, \tag{2.44a}$$

given in Eq. (2.37c), can be rewritten using Eq. (2.37b) as

$$\mathbf{H}_{F \text{ kin}} = \int d^3x \, \text{Tr} p_\psi \gamma^0 \vec{\gamma} \cdot \vec{\partial}\psi. \tag{2.44b}$$

Using the Majorana representation Dirac gamma matrices given in Appendix D, we see that in Majorana representation $\gamma^0\vec{\gamma}$ is a symmetric matrix in its spinor indices. Since the partial derivative $\vec{\partial}$ changes sign under integration by parts, we see that Eq. (2.44b) corresponds to the case of Eq. (2.43c), and thus in Majorana representation $\mathbf{H}_{F \text{ kin}}$ is even under interchange of p_F with q_F. When interactions are included, a similar analysis shows that

$$\mathbf{H}_{F \text{ int}} = \int d^3x \, \text{Tr}(g_S\overline{\psi}\{\phi_S, \psi\} + g_P\overline{\psi}i\gamma_5\{\phi_P, \psi\} + g_V\overline{\psi}i\vec{\gamma} \cdot [\vec{A}, \psi]), \tag{2.44c}$$

with ϕ_S, ϕ_P, and $\vec{A}$ respectively self-adjoint bosonic scalar, pseudoscalar, and vector matrix fields, gives an interaction trace Hamiltonian that is also even in Majorana representation under interchange of p_F and q_F. Thus the usual commutator form for a vector gauge field coupling, and symmetrized scalar and pseudoscalar field couplings, lead to a fermionic trace Hamiltonian $\mathbf{H}_F$ that is even in Majorana representation under interchange of fermionic canonical coordinates and momenta. Therefore, when the purely bosonic terms are included, we obtain in this case a total Hamiltonian $\mathbf{H}$ that is even under this interchange. We shall refer to these results in our statistical mechanical discussion later on.

3

Trace dynamics models with global supersymmetry*

In Section 2.4, we illustrated the trace dynamics formalism by constructing the trace dynamics analog of a simple field theory model, in which a Dirac fermion interacts with a scalar Klein–Gordon field. Much of the recent literature in quantum field theory has concerned itself with supersymmetric theories, in which invariance under the Poincaré group has been extended to invariance under the graded Poincaré group, and theories of this type are considered likely to play a central role in the ultimate unification of the forces. Our aim in this chapter (which can be omitted on a first reading) is to show that the trace dynamics formalism naturally extends to globally supersymmetric theories. Specifically, we shall see that, when there is a global supersymmetry, there is a conserved trace supersymmetry current with a time-independent trace supercharge $\mathbf{Q}_\alpha$, that together with the trace four momentum obeys the Poincaré supersymmetry algebra under the generalized Poisson bracket of Eq. (1.11a). We shall illustrate this statement with three concrete examples, the trace dynamics versions (Adler 1997a,b) of the Wess–Zumino model (Section 3.1), the supersymmetric Yang–Mills model (Section 3.2), and the so-called "matrix model for M theory" (Section 3.3). These three examples are worked out using component field methods; we close in Section 3.4 with a short discussion of a superspace approach, and of the obstruction that prevents the construction of a trace dynamics theory with local supersymmetry.

3.1 The Wess–Zumino model

We begin with the trace dynamics transcription of the Wess–Zumino model. We follow the notational conventions of West (1990), except that we normalize the fermion terms in the action differently, and always use the Majorana representation for the Dirac gamma matrices, in which $\gamma^{1,2,3}$ are real symmetric and $\gamma^0, i\gamma^5$ are real skew-symmetric. (For useful cyclic identities satisfied by this representation of the γ matrices, see Appendix D.) The trace Lagrangian for the Wess–Zumino model is

$$\mathbf{L} = \int d^3x \operatorname{Tr} \left(-\frac{1}{2}(\partial_\mu A)^2 - \frac{1}{2}(\partial_\mu B)^2 - \bar{\chi}\gamma^\mu \partial_\mu \chi + \frac{1}{2}F^2 + \frac{1}{2}G^2 \right.$$
$$- m(AF + BG - \bar{\chi}\chi)$$
$$\left. - \lambda[(A^2 - B^2)F + G\{A, B\} - 2\bar{\chi}(A - i\gamma_5 B)\chi] \right), \tag{3.1}$$

with A, B, F, G self-adjoint $N \times N$ Lorentz scalar matrices and with χ a Grassmann four-component column vector spinor, each spin component of which is a self-adjoint Grassmann $N \times N$ matrix. The notation $\bar{\chi}$ is defined by $\bar{\chi} = \chi^T\gamma^0$, with the transpose T acting only on the Dirac spinor structure, so that χ^T is the four-component row vector spinor constructed from the same $N \times N$ matrices that appear in χ. The numerical parameters λ and m are respectively the coupling constant and mass. Equation (3.1) is identical in appearance to the usual Wess–Zumino model Lagrangian, except that we have explicitly symmetrized the term $G\{A, B\}$; symmetrization of the other terms is automatic (up to total derivatives that do not contribute to the action) by virtue of the cyclic property of the trace.

Taking operator variations of Eq. (3.1) by using the recipe of Eq. (1.3b), the Euler–Lagrange equations of Eq. (1.7c) take the form

$$\partial^2 A = mF + \lambda(\{A, F\} + \{B, G\} - 2\bar{\chi}\chi),$$
$$\partial^2 B = mG + \lambda(-\{B, F\} + \{A, G\} + 2i\bar{\chi}\gamma_5\chi),$$
$$\gamma^\mu \partial_\mu \chi = m\chi + \lambda(\{A, \chi\} - i\{B, \gamma_5\chi\}), \tag{3.2}$$
$$F = mA + \lambda(A^2 - B^2),$$
$$G = mB + \lambda\{A, B\}.$$

Transforming to Hamiltonian form, the canonical momenta of Eq. (1.9a) are

$$p_\chi = -\bar{\chi}\gamma^0 = \chi^T,$$
$$p_A = \partial_0 A, \tag{3.3}$$
$$p_B = \partial_0 B,$$

and the trace Hamiltonian is given by

$$\mathbf{H} = \int d^3x \operatorname{Tr} \left(\frac{1}{2}[p_A^2 + p_B^2 + (\vec{\nabla}A)^2 + (\vec{\nabla}B)^2] + p_\chi\gamma^0\vec{\gamma} \cdot \vec{\nabla}\chi \right.$$
$$\left. + \frac{1}{2}(F^2 + G^2) - m\bar{\chi}\chi - \lambda p_\chi\gamma^0\{A - i\gamma_5 B, \chi\} \right), \tag{3.4a}$$

in which F and G are understood to be the functions of A and B given by the final two lines of Eq. (3.2), and where we have taken care to write $\mathbf{H}$ so that it is manifestly symmetric in the identical quantities p_χ and χ^T. The trace three-momentum

$\vec{\mathbf{P}}$, which together with $\mathbf{H}$ forms the trace four-momentum $\mathbf{P}^{\sigma}$ of Eq. (2.29c), is given by

$$\vec{\mathbf{P}} = -\int d^3x \operatorname{Tr}(p_A \vec{\nabla} A + p_B \vec{\nabla} B + p_\chi \vec{\nabla} \chi), \tag{3.4b}$$

while the conserved trace quantity $\mathbf{N}$ of Eq. (2.4e) and the conserved operator $\tilde{C}$ of Eq. (2.6) are given respectively by

$$\mathbf{N} = -i \int d^3x \operatorname{Tr} \chi^T \chi,$$

$$\tilde{C} = \int d^3x ([A, p_A] + [B, p_B] - \{\chi, p_\chi\}), \tag{3.5a}$$

with a contraction of the spinor indices in the final term on the second line of Eq. (3.5a) understood. The corresponding expressions for the conserved currents $\mathbf{N}^\mu$ and $\tilde{C}^\mu$, of which $\mathbf{N}$ and $\tilde{C}$ are the charges, are

$$\mathbf{N}^\mu = i \operatorname{Tr} \overline{\chi} \gamma^\mu \chi,$$

$$\tilde{C}^\mu = -[A, \partial^\mu A] - [B, \partial^\mu B] + \overline{\chi} \gamma^\mu \chi + \chi^T (\overline{\chi} \gamma^\mu)^T \tag{3.5b}$$

$$= -[A, \partial^\mu A] - [B, \partial^\mu B] + 2\overline{\chi} \gamma^\mu \chi,$$

with T indicating a transpose acting only on the Dirac spinor indices, and where the simplification leading to the final line is possible because χ is a Majorana spinor. Equations (3.4a,b) are clearly formed from the usual field-theoretic expressions for the Hamiltonian and three-momentum by taking the trace, and symmetrizing factors where this is not already implicit from the cyclic properties of the trace. Exactly the same procedure can be used to form the full trace energy-momentum tensor density $\mathcal{T}^{\mu\sigma}$.

Let us now perform a supersymmetry variation of the fields given by

$$\delta A = \overline{\epsilon} \chi, \quad \delta B = i \overline{\epsilon} \gamma_5 \chi,$$

$$\delta \chi = \frac{1}{2} [F + i \gamma_5 G + \gamma^\mu \partial_\mu (A + i \gamma_5 B)] \epsilon, \tag{3.6}$$

$$\delta F = \overline{\epsilon} \gamma^\mu \partial_\mu \chi, \quad \delta G = i \overline{\epsilon} \gamma_5 \gamma^\mu \partial_\mu \chi,$$

with ϵ a c-number Grassmann spinor (i.e., a four-component spinor, the spin components of which are 1×1 Grassmann matrices). Substituting Eq. (3.6) into the trace Lagrangian of Eq. (3.1), a lengthy calculation shows that when ϵ is constant, the variation of $\mathbf{L}$ vanishes. The calculation parallels that done in the conventional c-number Lagrangian case, except that the cyclic properties of the trace and cyclic identities obeyed by the Majorana representation γ matrices (see Appendix D) are used extensively in place of commutativity/anticommutativity of the fields. When

ϵ is not constant, the variation of **L** is given by

$$\delta\mathbf{L} = \int d^3x \mathrm{Tr}(\bar{J}^\mu \partial_\mu \epsilon),$$

$$\bar{J}^\mu = -\bar{\chi}\gamma^\mu\left[(\gamma^\nu\partial_\nu + m)(A + i\gamma_5 B) + \lambda(A^2 - B^2 + i\gamma_5\{A, B\})\right], \quad (3.7a)$$

which identifies the trace supercharge $\mathbf{Q}_\alpha$ as

$$\mathbf{Q}_\alpha \equiv \int d^3x \mathrm{Tr}\,\bar{J}^0\alpha$$

$$= \int d^3x \mathrm{Tr}\frac{1}{2}(p_\chi + \chi^T)\left[(\gamma^\nu\partial_\nu + m)(A + i\gamma_5 B) + \lambda(A^2 - B^2 + i\gamma_5\{A, B\})\right]\alpha. \quad (3.7b)$$

Here α is a constant auxiliary c-number Grassmann spinor that has been inserted to make the argument of the trace a bosonic quantity, and we have again taken care to express $\mathbf{Q}_\alpha$ symmetrically in the identical quantities p_χ and χ^T. It is straightforward to check, using the equations of motion and the cyclic identity, that $\bar{\mathbf{J}}^\mu = \mathrm{Tr}\bar{J}^\mu$ is a conserved trace supercurrent, which implies that the trace supercharge is time independent.

It is now straightforward (but tedious) to check the closure of the supersymmetry algebra under the generalized Poisson bracket of Eq. (1.11a), which for the Hamiltonian dynamics of the Wess–Zumino model gives

$$\{\mathbf{Q}_\alpha, \mathbf{Q}_\beta\} = \mathrm{Tr}\left[\frac{\delta\mathbf{Q}_\alpha}{\delta A}\frac{\delta\mathbf{Q}_\beta}{\delta p_A} + \frac{\delta\mathbf{Q}_\alpha}{\delta B}\frac{\delta\mathbf{Q}_\beta}{\delta p_B} - \sum_{d=1}^4 \frac{\delta\mathbf{Q}_\alpha}{\delta\chi^d}\frac{\delta\mathbf{Q}_\beta}{\delta p_{\chi^d}} - (\alpha \leftrightarrow \beta)\right]$$

$$= \bar{\alpha}\gamma^0\beta\mathbf{H} - \bar{\alpha}\vec{\gamma}\beta \cdot \vec{\mathbf{P}}, \quad (3.8a)$$

with $\mathbf{H}$ and $\vec{\mathbf{P}}$ the trace Hamiltonian and three-momentum given above. It is also easy to check that $\mathbf{Q}_\epsilon$ plays the role of the generator of supersymmetry transformations for the dynamical variables A, B, χ under the generalized Poisson bracket, since we readily find (for constant Grassmann even parameters a, b and Grassmann odd spinor parameter c)

$$\{\mathrm{Tr}(aA + bB + c^T\chi), \mathbf{Q}_\epsilon\} = \mathrm{Tr}(a\delta A + b\delta B + c^T\delta\chi), \quad (3.8b)$$

with $\delta A, \delta B, \delta\chi$ the supersymmetry variations given by Eq. (3.6) above, after elimination of the auxiliary fields F, G by their equations of motion.

3.2 The supersymmetric Yang–Mills model

As a second example of a trace dynamics model with global supersymmetry, we discuss supersymmetric Yang–Mills theory. We start from the trace

Lagrangian

$$\mathbf{L} = \int d^3x \mathrm{Tr}\left[\frac{1}{4g^2}F_{\mu\nu}^2 - \bar{\chi}\gamma^\mu D_\mu \chi + \frac{1}{2}D^2\right], \tag{3.9a}$$

with the field strength $F_{\mu\nu}$ and covariant derivative D_μ constructed from the gauge potential A_μ according to

$$F_{\mu\nu} = \partial_\mu A_\nu - \partial_\nu A_\mu + [A_\mu, A_\nu],$$
$$D_\mu \mathcal{O} = \partial_\mu \mathcal{O} + [A_\mu, \mathcal{O}] \tag{3.9b}$$
$$\Rightarrow D_\mu F_{\nu\lambda} + D_\nu F_{\lambda\mu} + D_\lambda F_{\mu\nu} = 0.$$

In Eqs. (3.9a,b), the potential components A_μ are each an anti-self-adjoint, and the auxiliary field D is a self-adjoint, $N \times N$ matrix, and each spinor component of χ is a self-adjoint Grassmann $N \times N$ matrix. The Euler–Lagrange equations of motion are

$$D = 0,$$
$$\gamma^\mu D_\mu \chi = 0, \tag{3.10a}$$
$$D_\mu F^{\mu\nu} = 2g^2 \bar{\chi}\gamma^\nu \chi;$$

as usual for a gauge system, the $\nu = 0$ component of Eq. (3.10a) is not a dynamical evolution equation, but rather the constraint

$$D_\ell F^{\ell 0} = 2g^2 \bar{\chi}\gamma^0 \chi. \tag{3.10b}$$

Going over to the Hamiltonian formalism, the canonical momenta are given by

$$p_{A_\ell} = -\frac{1}{g^2}F_{0\ell}, \quad p_\chi = \chi^T, \tag{3.11a}$$

and the axial gauge trace Hamiltonian is

$$\mathbf{H} = \mathbf{H}_A + \mathbf{H}_\chi, \tag{3.11b}$$

with

$$\mathbf{H}_A = \int d^3x \mathrm{Tr}\left(\frac{-g^2}{2}\sum_{\ell=1}^{2}p_{A_\ell}^2 - \frac{1}{2g^2}F_{03}^2 \right.$$
$$\left. -\frac{1}{2g^2}(\partial_1 A_2 - \partial_2 A_1 + [A_1, A_2])^2 - \frac{1}{2g^2}[(\partial_3 A_1)^2 + (\partial_3 A_2)^2]\right), \tag{3.11c}$$

$$F_{03} = \frac{1}{2}g^2 \int_{-\infty}^{\infty} dz' \epsilon(z - z')[-(p_\chi \chi + \chi^T p_\chi^T) + D_1 p_{A_1} + D_2 p_{A_2}]|_{z'},$$

$$\mathbf{H}_\chi = \int d^3x \mathrm{Tr}(p_\chi \gamma^0 \gamma_\ell D_\ell \chi),$$

where we have taken care to write **H** in a form symmetric in the identical quantities

p_χ and χ^T, and where $\epsilon(z) = 1(-1)$ for $z > 0(z < 0)$. The trace three-momentum is

$$\mathbf{P}_m = -\int d^3x \, \mathrm{Tr}\left(\sum_{\ell=1}^{3} F_{m\ell} \, p_{A_\ell} + p_\chi D_m \chi\right), \tag{3.12}$$

and the conserved operator $\tilde{C}$ of Eq. (2.6) is given by

$$\tilde{C} = \int d^3x \left(\sum_{\ell=1}^{2} [A_\ell, p_{A_\ell}] - \{\chi, p_\chi\}\right), \tag{3.13a}$$

with a contraction of the spinor indices in the final term of Eq. (3.13a) understood. By virtue of the constraint of Eq. (3.10b), the conserved operator $\tilde{C}$ can also be written as

$$\tilde{C} = -\int d^3x \sum_{\ell=1}^{3} \partial_\ell p_{A_\ell} = -\int_{\text{sphere at } \infty} d^2 S_\ell \, p_{A_\ell}, \tag{3.13b}$$

which vanishes when the surface integral in Eq. (3.13b) is zero. The corresponding conserved current $\tilde{C}^\mu$, of which $\tilde{C}$ is the charge, is given by

$$\tilde{C}^\mu = \frac{1}{g^2}[A_\nu, F^{\mu\nu}] + 2\overline{\chi}\gamma^\mu \chi. \tag{3.13c}$$

The conserved trace quantity $\mathbf{N}$ and the corresponding conserved current $\mathbf{N}^\mu$ have the same form as in the Wess–Zumino model

$$\mathbf{N} = -i \int d^3x \, \mathrm{Tr}\chi^T \chi,$$

$$\mathbf{N}^\mu = i \, \mathrm{Tr}\overline{\chi}\gamma^\mu \chi. \tag{3.14}$$

Making now the supersymmetry variations

$$\delta A_\mu = ig\overline{\epsilon}\gamma_\mu \chi,$$

$$\delta \chi = \left(\frac{i}{8g}[\gamma_\mu, \gamma_\nu]F^{\mu\nu} + \frac{i}{2}\gamma_5 D\right)\epsilon, \tag{3.15}$$

$$\delta D = i\overline{\epsilon}\gamma_5\gamma^\mu D_\mu \chi,$$

in the trace Lagrangian, with ϵ again a c-number Grassmann spinor, we find using cyclic invariance under the trace and the γ matrix identities given in Appendix D that when ϵ is constant, the variation vanishes. When ϵ is not a constant, the variation of $\mathbf{L}$ is given by

$$\delta\mathbf{L} = \int d^3x \, \mathrm{Tr}(\bar{J}^\mu \partial_\mu \epsilon),$$

$$\bar{J}^\mu = -\frac{i}{4g}\overline{\chi}\gamma^\mu F_{\nu\sigma}[\gamma^\nu, \gamma^\sigma], \tag{3.16a}$$

from which we construct the trace supercharge $\mathbf{Q}_\alpha$ as

$$\mathbf{Q}_\alpha = \int d^3x \, \mathrm{Tr} \frac{i}{8g} (p_\chi + \chi^T) F_{\nu\sigma} [\gamma^\nu, \gamma^\sigma] \alpha, \qquad (3.16b)$$

with α a constant c-number Grassmann spinor. Again, it is straightforward to check, using the equations of motion and the cyclic identity, that $\bar{J}_\mu = \mathrm{Tr} \bar{J}_\mu$ is a conserved trace supercurrent, which implies that the trace supercharge is conserved.

One can now verify the closure of the supersymmetry algebra under the generalized Poisson bracket of Eq. (1.11a), which for the Hamiltonian dynamics of the supersymmetric Yang–Mills model gives

$$\{\mathbf{Q}_\alpha, \mathbf{Q}_\beta\} = \mathrm{Tr} \left[\sum_{l=1}^{2} \frac{\delta \mathbf{Q}_\alpha}{\delta A_\ell} \frac{\delta \mathbf{Q}_\beta}{\delta p_{A_\ell}} - \sum_{d=1}^{4} \frac{\delta \mathbf{Q}_\alpha}{\delta \chi^d} \frac{\delta \mathbf{Q}_\beta}{\delta p_{\chi^d}} - (\alpha \leftrightarrow \beta) \right]$$
$$= \bar{\alpha} \gamma^0 \beta \mathbf{H} - \bar{\alpha} \vec{\gamma} \beta \cdot \vec{\mathbf{P}}, \qquad (3.17)$$

with $\mathbf{H}$ and $\vec{\mathbf{P}}$ given by Eqs. (3.11b,c) and Eq. (3.12) respectively. Examining the role of the supercharge as a generator of transformations, in analogy with Eq. (3.8b), the supercharge in the Yang–Mills case is found to generate the supersymmetry variations of Eq. (3.15), plus an infinitesimal change of gauge.

3.3 The matrix model for M theory

As our third example of a trace dynamics model with global supersymmetry, we consider the matrix model studied by Bergshoeff, Sezgin, and Townsend (1987, 1988), Claudson and Halpern (1985), Rittenberg and Yankielowicz (1985), and Flume (1985), surveyed in de Wit (1997). This has recently been studied by de Wit, Hoppe, and Nicolai (1988), Townsend (1996), Banks *et al.* (1997), and Banks, Seiberg, and Shenker (1997) in a string-theory context under the name "the matrix model for M theory"; for a review see Taylor (2001). This model, formulated in zero spatial dimensions, has the trace Lagrangian $\mathbf{L}$ given by

$$\mathbf{L} = \mathrm{Tr} \left(\frac{1}{2} D_t X_i D_t X^i + \theta^T D_t \theta + \frac{1}{4} [X_i, X_j][X^i, X^j] - i\theta^T \gamma_i [\theta, X^i] \right), \qquad (3.18)$$

with the covariant derivative defined now by $D_t \mathcal{O} = \partial_t \mathcal{O} - i[A_0, \mathcal{O}]$. In Eq. (3.18), a summation convention is understood on the indices i, j which range from 1 to 9; A_0 and the X_i are self-adjoint $N \times N$ complex matrices, while θ is a 16-component fermionic spinor each element of which is a self-adjoint $N \times N$ complex Grassmann matrix, with the transpose T acting only on the spinor structure but not on the $N \times N$ matrices, so that θ^T is simply the 16-component row spinor corresponding to the 16-component column spinor θ. The potential A_0 has no kinetic term and so is a pure gauge degree of freedom. Finally, the γ_i are a set of nine 16×16 matrices, which are related to the standard 32×32 matrices Γ^μ as well as to the Dirac matrices of spin(8), as conveniently described in Green, Schwartz, and Witten (1987) and obeying identities summarized in Appendix D.

Starting from the trace Lagrangian of Eq. (3.18), using the definition of Eq. (1.3b) to take operator variations, the operator Euler–Lagrange equations of Eq. (1.7c) give the equations of motion of the matrix model

$$D_t^2 X^i = [[X^j, X^i], X_j] - i2\theta^T \gamma^i \theta,$$
$$D_t \theta^T = i[\theta^T \gamma_i, X^i] \implies D_t \theta = i[\gamma_i \theta, X^i], \tag{3.19a}$$

together with the constraint that the generic conserved operator $\tilde{C}$ of Eq. (2.6) vanishes in this model

$$\tilde{C} = [X^i, D_t X_i] - 2\theta^T \theta = 0. \tag{3.19b}$$

The vanishing of $\tilde{C}$ here is an analog of the fact that in the supersymmetric Yang–Mills model of Section 3.2, we found that $\tilde{C}$ is a surface integral at spatial infinity, and vanishes when this surface integral vanishes. The model of this section has the structure of a zero spatial dimension supersymmetric Yang–Mills theory, and so the surface integral at infinity of Section 3.2 is replaced here by zero.

To transform the dynamics to trace Hamiltonian form, we define the canonical momenta p_{X_i} and p_θ by

$$p_{X_i} = \frac{\delta \mathbf{L}}{\delta(\partial_t X_i)} = D_t X^i,$$

$$p_\theta = \frac{\delta \mathbf{L}}{\delta(\partial_t \theta)} = \theta^T, \tag{3.20a}$$

so that the trace Hamiltonian is given by

$$\mathbf{H} = \mathrm{Tr}(p_{X_i} \partial_t X_i + p_\theta \partial_t \theta) - \mathbf{L} = \mathrm{Tr}\left(\frac{1}{2} p_{X_i} p_{X^i} - \frac{1}{4}[X_i, X_j][X^i, X^j]\right.$$

$$\left. + i p_\theta \gamma_i [\theta, X^i] + i A_0 \tilde{C}\right). \tag{3.20b}$$

Again, because $p_\theta = \theta^T$, we have written the trace Hamiltonian in a form that is manifestly symmetric under the replacements $p_\theta \to \theta^T$, $\theta \to p_\theta^T$.

Let us next consider the variation of the trace Lagrangian under the supersymmetry transformation defined by

$$\delta X^i = 2i\epsilon^T \gamma^i \theta = -2i\theta^T \gamma^i \epsilon,$$

$$\delta\theta = -\left(i D_t X^i \gamma_i + \frac{1}{2}[X^i, X^j]\gamma_{ij}\right)\epsilon + \epsilon', \tag{3.21}$$

$$\delta A_0 = 2i\epsilon^T \theta = -2i\theta^T \epsilon.$$

Here ϵ and ϵ' are 16-component Grassmann c-number spinors, that is, they are column vectors each of whose 16 components is an independent 1×1 Grassmann matrix. Using the cyclic trace identities of Section 1.1 and the γ matrix properties summarized in Appendix D, it is a matter of straightforward but lengthy calculation to verify that the trace Lagrangian is invariant under the transformation of Eq. (3.21) when ϵ and ϵ' are time independent. When ϵ and ϵ' have a time dependence, $\delta \mathbf{L}$ is no longer zero, but instead is given by

$$\delta \mathbf{L} = \partial_t \mathrm{Tr} \left(-\theta^T \epsilon' - (i\theta^T \gamma_i D_t X^i - \frac{1}{2}\theta^T \gamma_{ij}[X^i, X^j])\epsilon \right)$$

$$+ \mathrm{Tr}(2\theta^T \partial_t \epsilon' - (2i\theta^T \gamma_i D_t X^i + \theta^T \gamma_{ij}[X^i, X^j])\partial_t \epsilon). \quad (3.22a)$$

This identifies the trace supercharges $\mathbf{Q}'_\alpha$ and $\mathbf{Q}_\alpha$ as

$$\mathbf{Q}'_\alpha = \mathrm{Tr} 2\theta^T \alpha,$$
$$\mathbf{Q}_\alpha = \mathrm{Tr}\left(- 2i\theta^T \gamma_i D_t X_i - \theta^T \gamma_{ij}[X^i, X^j]\right)\alpha, \quad (3.22b)$$

with α a 16-component c-number Grassmann spinor, and their conservation is easily checked using the equations of motion and γ matrix identities. To check the supersymmetry algebra, we must first write the supercharges of Eq. (3.22b) in Hamiltonian form, symmetrized with respect to p_θ and $i\theta^T$, giving

$$\mathbf{Q}'_\alpha = \mathrm{Tr}(p_\theta + \theta^T)\alpha,$$
$$\mathbf{Q}_\alpha = -\mathrm{Tr}(p_\theta + \theta^T)\left(i\gamma_i p_{X_i} + \frac{1}{2}\gamma_{ij}[X^i, X^j]\right)\alpha. \quad (3.22c)$$

Using the generalized Poisson bracket corresponding to the Hamiltonian structure of our model, defined now by

$$\{\mathbf{A}, \mathbf{B}\} = \mathrm{Tr}\left[\frac{\delta \mathbf{A}}{\delta X_i}\frac{\delta \mathbf{B}}{\delta p_{X_i}} - \frac{\delta \mathbf{B}}{\delta X_i}\frac{\delta \mathbf{A}}{\delta p_{X_i}} - \sum_{d=1}^{16}\left(\frac{\delta \mathbf{A}}{\delta \theta^d}\frac{\delta \mathbf{B}}{\delta p_{\theta d}} - \frac{\delta \mathbf{B}}{\delta \theta^d}\frac{\delta \mathbf{A}}{\delta p_{\theta d}}\right)\right], \quad (3.23)$$

it is straightforward to evaluate the supercharge algebra, and to show that it has the expected form (Adler, 1997b).

3.4 Superspace considerations and remarks

The derivations of Sections 3.1, 3.2, and 3.3 have all been carried out in the component formalism, which requires doing a separate computation for each Poincaré supersymmetry multiplet. However, there is a simple and general superspace argument for the results we have obtained. Recall that superspace is constructed by introducing four fermionic coordinates θ_α corresponding to the four spacetime coordinates x_μ (see, e.g., West, 1990, Chapter 14). The graded Poincaré algebra is then represented by differential operators constructed from the superspace coordinates, and superfields are represented by finite polynomials in the fermionic coordinates θ_α, with coefficient functions that depend on x_μ. To generalize the superspace formulation to give trace dynamics models, one simply replaces these coefficient functions by $N \times N$ matrices (or operators), and inserts a trace Tr acting on the superspace integrals used to form the action. Then the standard argument, that the action is invariant under superspace translations of the x_μ and the θ_α, still holds for the trace action formed this way from the matrix components of the superfields. We immediately see from this argument why it is essential for the supersymmetry parameter ϵ to be a Grassmann c-number and not also a matrix; this parameter appears as the magnitude of an infinitesimal superspace translation, and, since the superspace coordinates x_μ and θ_α are c-numbers, the parameter ϵ must be one also. The construction just given gives reducible supersymmetry representations, and various constraints, constructed from differential operators with Dirac γ matrix coefficients acting on the superspace coordinates, must be applied to the superfields to pick out irreducible representations. Since these constraints act linearly on the expansion coefficients, and involve no non-commutative operators, they can all be immediately generalized (with the usual replacement of complex conjugation for c-numbers by the adjoint) to the case in which the coefficient functions are matrices or operators.

The simplicity of this argument suggests that generally, for rigid supersymmetry theories for which there exists a superspace construction, there will exist a corresponding trace dynamics generalization. The superspace argument also suggests why it has not been possible (unpublished investigation by the author and Y.-S. Wu) to construct trace dynamics generalizations of local supersymmetry theories, such as supergravity. The commutator of two local supersymmetries with supersymmetry parameters ϵ_1 and ϵ_2 is a linear combination (see, e.g., West, 1990, Chapter 9) of a local Lorentz transformation, a general coordinate transformation, and a supersymmetry transformation with supersymmetry parameter proportional to

$$\bar{\epsilon}_2 \gamma_\mu \epsilon_1 \psi^\mu, \qquad (3.24)$$

with ψ^μ the Rarita–Schwinger gravitino field. Even if we start with $\epsilon_{1,2}$ that are c-numbers, the new supersymmetry parameter given by Eq. (3.24) will be matrix valued in a trace dynamics generalization where the gravitino field ψ^μ is matrix valued. Thus, an extension of the results of this chapter to local supersymmetries would appear to require a generalization of the results presented above to the case in which the supersymmetry parameter ϵ is matrix-valued, rather than a c-number as assumed throughout our discussion. If, as we suspect, such a generalization is not possible, there will be implications for the strategy to be followed in incorporating gravity and supergravity into a trace dynamics framework. For example, perhaps the gravitational degrees of freedom should be treated not as matrix quantities but only as c-number classical ones, a possibility consistent with the view, discussed for example by Weinstein (2001), that perhaps gravity should not be "quantized" in the same manner that other fields are. Another (not mutually exclusive) possibility is that perhaps the gravitational degrees of freedom are emergent degrees of freedom arising from underlying degrees of freedom, with the underlying degrees of freedom obeying a rigid supersymmetry with a trace dynamics generalization.

4

Statistical mechanics of matrix models

Up to this point we have discussed matrix models as classical dynamical systems. We shall now start laying the groundwork for the emergence of quantum mechanical behavior from a matrix dynamics in which quantization is not assumed a priori. We begin this discussion by emphasizing that we shall *not* follow the traditional route (see Brézin and Wadia, 1993) of canonically quantizing a matrix model, in which working from Eqs. (1.8) and (1.9a) one takes each classically conjugate matrix element pair $(q_r)_{ij}$ and $(p_r)_{ji} = \partial L / \partial (\dot{q}_r)_{ij}$, and elevates them to quantum operators that satisfy canonical commutation relations, such as (for bosonic degrees of freedom)

$$[(q_r)_{ij}, (q_s)_{kl}] = [(p_r)_{ij}, (p_s)_{kl}] = 0,$$
$$[(q_r)_{ij}, (p_s)_{kl}] = i\hbar \delta_{rs} \delta_{il} \delta_{jk}. \tag{4.1}$$

In this approach, each classical matrix q_r, because it has N^2 matrix elements, ends up spawning N^2 quantum operators. The canonical quantization approach is appropriate, for example, when dealing with a matrix model that arises as a discretized approximation to a continuum field system, or as an approximation to a many-body system with a large number of independent degrees of freedom. In these cases, the matrix elements each represent a degree of freedom to which, assuming that one is dealing with a quantum field or a quantum many-body system, the usual quantization rules apply. To repeat, this is *not* what we shall do, because we are not assuming that quantum theory applies at the underlying trace dynamics level.

Instead, we shall require that the only matrix (or operator) structure present is that which is already present in the classical matrix model (and, in this aspect, our approach shares a common philosophy with that of Smolin, 1983, 1985, 2002). Thus, each q_r and its associated p_r correspond to a single operator degree of freedom, and these degrees of freedom do not obey any simple commutation algebra: in general for bosonic degrees of freedom, $[q_r, q_s]$, $[p_r, p_s]$, and $[q_r, p_s]$ will all be non-zero, and similarly when fermionic degrees of freedom are included, with anticommutators replacing commutators as appropriate. However, we shall argue in Chapter 5 that in the statistical dynamics of matrix models with a global unitary invariance, within thermodynamic averages over polynomials of the qs and ps, the matrix variables q_r and p_s obey (when specific approximations are made) an *effective* commutator/anticommutator algebra of the familiar canonical form. In other words, we shall show that quantum behavior is an emergent feature of the statistical mechanics of a particular class of matrix models, with each matrix variable pair q_r, p_r corresponding to *one* quantum mechanical operator degree of freedom.

The first step in such a program is to set up the statistical mechanics of matrix models, and that is what we shall do in this chapter. The basic prerequisite for applying statistical mechanical methods is to establish a Liouville theorem, and this is addressed in Section 4.1, where we show that the natural matrix phase space integration measure is invariant under general canonical transformations. Next, in Section 4.2, we derive the canonical ensemble for trace dynamics, by maximizing the entropy subject to the constraints imposed by the generic conserved quantities established in Chapters 1 and 2. We also introduce here an assumption that will be used in the subsequent analysis of Chapter 5, that the ensemble does not single out any special states in the underlying Hilbert space. In Section 4.3, we give an alternative derivation of the canonical ensemble, by starting from the microcanonical ensemble. In Section 4.4, which can be skipped on a first reading, we give the analogs of standard gauge fixing methods that are needed to formulate the canonical ensemble for trace dynamics theories with a local gauge invariance, such as the Yang–Mills model discussed in Section 3.2. In Section 4.5 we continue the discussion of canonical ensemble averages begun in Section 4.2, by studying the implications of the fact that the canonical ensemble leaves unbroken a subgroup of the originally assumed global unitary invariance group. This necessitates a unitary fixing procedure for the canonical ensemble. The technical details of the unitary fixing are given in Section 4.6, which can also be skipped on a first reading.

4.1 The Liouville theorem

We begin our statistical treatment of matrix models by deriving, following Adler and Millard (1996), an analog of the Liouville theorem, which states that the matrix model trace dynamics leaves a suitably defined phase space volume element invariant. We shall actually derive a more general result, showing that the phase space volume element is invariant under the general canonical transformations introduced in Eqs. (2.13a,b), of which evolution under the trace Hamiltonian dynamics, and unitary transformation of the Hilbert space bases, are both special cases.

Following the notation of Section 1.1, we denote the general matrix element of the operator x_r by $(x_r)_{mn}$, which can be decomposed into real and imaginary parts according to

$$(x_r)_{mn} = (x_r)^0_{mn} + i(x_r)^1_{mn}, \tag{4.2a}$$

where $(x_r)^A_{mn}$ with $A = 0, 1$ are real numbers. If for the moment we ignore adjointness restrictions, the natural phase space measure is defined by

$$d\mu = \prod_A d\mu^A,$$

$$d\mu^A \equiv \prod_{r,m,n} d(x_r)^A_{mn}; \tag{4.2b}$$

when adjointness restrictions are taken into account, certain factors in Eq. (4.2b) become redundant and are omitted. For bosonic operators x_r, we shall assume that the matrix element phase space is unbounded, so that the limits for bosonic phase space integrations with the measure of Eq. (4.2b) are $-\infty < (x_r)_{mn} < \infty$. For fermionic operators, Grassmann integration is effectively differentiation, and we assume that the phase space is again unbounded in the sense that the usual Grassmann integration formulas, which are invariant under a constant translation of the integration variable, are valid. (See the introductory paragraphs to the Appendices for further details on Grassmann integrals and their manipulation, and references.) Our strategy is first to ignore adjointness restrictions and to prove the canonical invariance of each individual factor $d\mu^A$ in the first line of Eq. (4.2b), and then to indicate how the argument is altered when adjointness restrictions are taken into account.

Under the general canonical transformation of Eq. (2.13b), the matrix elements of the new variables $\hat{x}_r \equiv x_r + \delta x_r$ are related to those of the original variables x_r by

$$(\hat{x}_r)^A_{mn} = (x_r)^A_{mn} + \sum_s \omega_{rs} \left(\frac{\delta G}{\delta x_s} \right)^A_{mn}. \tag{4.3a}$$

Inserting a complete set of intermediate states into the fundamental definition

$$\delta G = \text{Tr} \sum_s \frac{\delta G}{\delta x_s} \delta x_s, \tag{4.3b}$$

and using the reality of G, we get

$$\delta G = \sum_{s,m,n,A} \epsilon^A \left(\frac{\delta G}{\delta x_s} \right)^A_{mn} (\delta x_s)^A_{nm}, \tag{4.3c}$$

where $\epsilon^0 = 1$ and $\epsilon^1 = -1$. Thus, we see that

$$\left(\frac{\delta \mathbf{G}}{\delta x_s}\right)^A_{mn} = \epsilon^A \frac{\partial \mathbf{G}}{\partial (x_s)^A_{nm}}, \tag{4.3d}$$

a result that can also be obtained by decomposing Eq. (1.8) into real and imaginary parts (with the factor ϵ^A then arising from the facts that $1^{-1} = 1$ while $i^{-1} = -i$). Equation (4.3d) allows us to rewrite Eq. (4.3a) in terms of conventional partial derivatives of the trace functional $\mathbf{G}$

$$(\hat{x}_r)^A_{mn} = (x_r)^A_{mn} + \sum_s \omega_{rs} \epsilon^A \frac{\partial \mathbf{G}}{\partial (x_s)^A_{nm}}. \tag{4.3e}$$

Differentiating Eq. (4.3e) with respect to $(x_{r'})^A_{m'n'}$, we get for the transformation matrix

$$\frac{\partial (\hat{x}_r)^A_{mn}}{\partial (x_{r'})^A_{m'n'}} = \delta_{rr'} \delta_{mm'} \delta_{nn'} + \sum_s \omega_{rs} \epsilon^A \frac{\partial^2 \mathbf{G}}{\partial (x_s)^A_{nm} \partial (x_{r'})^A_{m'n'}}. \tag{4.4}$$

Since for an infinitesimal matrix δX we have $\det(1 + \delta X) \approx 1 + \mathrm{Tr}\delta X$, we learn from Eq. (4.4) that the Jacobian of the transformation is

$$J = 1 + \Sigma,$$

$$\Sigma = \sum_{r,s,m,n} \omega_{rs} \epsilon^A \frac{\partial^2 \mathbf{G}}{\partial (x_s)^A_{nm} \partial (x_r)^A_{mn}}. \tag{4.5a}$$

Interchanging in the expression for Σ in Eq. (4.5a) the summation indices r and s, and also interchanging the summation indices m and n, we get

$$\Sigma = \sum_{r,s,m,n} \omega_{sr} \epsilon^A \frac{\partial^2 \mathbf{G}}{\partial (x_r)^A_{mn} \partial (x_s)^A_{nm}}. \tag{4.5b}$$

However, now using the fact (cf. Eq. (1.17)) that for bosonic r, s we have

$$\omega_{sr} = -\omega_{rs},$$

$$\frac{\partial^2 \mathbf{G}}{\partial (x_r)^A_{mn} \partial (x_s)^A_{nm}} = \frac{\partial^2 \mathbf{G}}{\partial (x_s)^A_{nm} \partial (x_r)^A_{mn}}, \tag{4.5c}$$

while for fermionic r, s we have

$$\omega_{sr} = \omega_{rs},$$

$$\frac{\partial^2 \mathbf{G}}{\partial (x_r)^A_{mn} \partial (x_s)^A_{nm}} = -\frac{\partial^2 \mathbf{G}}{\partial (x_s)^A_{nm} \partial (x_r)^A_{mn}}, \tag{4.5d}$$

we see that Eqs. (4.5a–d) imply that $\Sigma = -\Sigma$; hence Σ vanishes and the Jacobian of the transformation is unity.

Now let us see how this argument is modified when we take the adjointness restrictions on the phase space variables x_r into account. Inspection of the argument just given shows that the diagonal $(m = n)$ and off-diagonal $(m \neq n)$ terms in the sum Σ vanish separately, and for each of these the summed contribution from the canonical coordinate and momentum pair q_r, p_r for each fixed r also vanishes separately. This observation permits us to readily take the adjointness restrictions into account; in the following discussion we shall write $d\mu = d\mu_B d\mu_F$, with $d\mu_B$ and $d\mu_F$ respectively the bosonic and fermionic integration measures.

For a bosonic pair of phase space variables q_r, p_r, the x_r variables are independent but are both self-adjoint or both anti-self-adjoint. We consider in detail the self-adjoint case, for which we have

$$(x_r)_{mn}^A = \epsilon^A (x_r)_{nm}^A. \tag{4.6a}$$

This means that the integration measure must be redefined to include only the factors that are real diagonal in m, n (the imaginary diagonal ones are identically zero), and only the upper diagonal off-diagonal factors (since the lower diagonal ones are related to the upper diagonal ones by complex conjugation), so that the bosonic integration measure becomes

$$d\mu_B = \prod_{r,m} d(x_r)_{mm}^0 \prod_{r,m<n,A} d(x_r)_{mn}^A. \tag{4.6b}$$

Equation (4.5b) is now replaced by

$$\Sigma = \sum_{r,s,m} \omega_{sr} \frac{\partial^2 G}{\partial (x_r)_{mm}^0 \partial (x_s)_{mm}^0} + \sum_{r,s,m<n,A} \omega_{sr} \frac{\partial^2 G}{\partial (x_r)_{mn}^A \partial (x_s)_{mn}^A}, \tag{4.6c}$$

both terms of which vanish by virtue of the antisymmetry of ω_{sr} in its indices in the bosonic sector. The redefinition of the measure in the anti-self-adjoint case proceeds similarly, with the replacement of $d(x_r)_{mm}^0$ by $d(x_r)_{mm}^1$ in Eq. (4.6b).

For a fermionic pair of phase space variables constructed according to the recipe $p_r = q_r^\dagger$ of Eq. (2.4b), the x_r variables are no longer independent. However, this construction implies that

$$(p_r)_{mn}^A = \epsilon^A (q_r)_{nm}^A, \tag{4.7a}$$

and thus the fermionic integration measure must be redefined as

$$d\mu_F = \prod_{r,m,n,A} d(q_r)_{mn}^A. \tag{4.7b}$$

But now we note that

$$d(q_r^0)_{mn}d(q_r^1)_{mn}d(q_r^0)_{nm}d(q_r^1)_{nm} \propto d(q_r^0 + iq_r^1)_{mn}d(q_r^0 - iq_r^1)_{mn}$$
$$\times d(q_r^0 + iq_r^1)_{nm}d(q_r^0 - iq_r^1)_{nm}$$
$$= d(q_r)_{mn}d(q_r^\dagger)_{mn}d(q_r)_{nm}d(q_r^\dagger)_{nm}$$
$$= d(q_r)_{mn}d(p_r)_{mn}d(q_r)_{nm}d(p_r)_{nm}, \qquad (4.7c)$$

and so the integration measure of Eq. (4.7b) can be rewritten as

$$d\mu_F = \prod_{r,m,n} d(x_r)_{mn}, \qquad (4.7d)$$

with the $d(x_r)_{mn}$ now complex differentials. The argument then proceeds just as in the unrestricted case, but with omission of the superscript A, since we recall that our original argument demonstrated the canonical invariance of the individual factors $d\mu^0$ and $d\mu^1$ in Eq. (4.2b).

An alternative procedure, that uses only real Grassmann differentials, is to note that for the measure of Eq. (4.7b), the sum over r in Eq. (4.5b) is restricted to only range over fermionic variables q_r, but not their conjugates p_r. Since ω_{sr} is block diagonal over the canonical variables, the index s must correspond to the variable p_r conjugate to q_r. Hence substituting Eq. (4.7a) into Eq. (4.5b), and using Eqs. (1.16a,b), we have

$$\Sigma = - \sum_{r,m,n,A} \frac{\partial^2 \mathbf{G}}{\partial(q_r)_{mn}^A \partial(q_r)_{mn}^A} = 0. \qquad (4.7e)$$

Finally, one can show that similar arguments holds for the more general fermionic adjointness assignment of Eqs. (2.17a–d), provided that A_{rs} is such that $\det \mathcal{A}_{\alpha\beta} \neq 0$, where $\mathcal{A}_{\alpha\beta} = (A_{rs})_{mn}$, with α and β the composite row and column indices given by $\alpha = (r, m)$ and $\beta = (s, n)$.

To summarize, we have shown that the matrix operator phase space integration measure $d\mu$ is invariant under general canonical transformations. As noted at the beginning of this section, an important corollary of this result follows when $\mathbf{G}$ is taken as the generator $dt\mathbf{H}$ of an infinitesimal time translation, since we then learn that $d\mu$ is invariant under the dynamical evolution of the system, giving a trace dynamics analog of Liouville's theorem of classical mechanics. Since no restrictions on the form of the generator $\mathbf{G}$ were needed in the above argument for the invariance of $d\mu$, the argument applies even when $\mathbf{G}$ is formed from the operator phase space variables using *operator* coefficients. Thus, the integration measure $d\mu$ is invariant under a unitary transformation on the basis of states in Hilbert space, the effect of which on the variables $\{x_r\}$ can be represented by Eqs. (2.11a–c). (Note, however, that this transformation is not itself global unitary invariant (cf. Eq. (2.12a)), and so is only a covariance, rather than an invariance, of the conserved operator $\tilde{C}$.)

4.2 The canonical ensemble

The matrix equations of motion of trace dynamics determine the time evolution of the matrix qs and ps at all times, given their values on an initial time slice. However, these initial values are themselves not determined. We shall now make the assumption that for a large enough system, the phase space distribution of the matrix variables rapidly loses memory of fine details of their initial values, and that over relevant experimental resolution times the system uniformly samples all phase space configurations that are consistent with the generic conservation laws. This allows us to represent time averages of physical quantities over experimental resolution times as averages at a fixed time, taken over an equilibrium ensemble representing the different phase space configurations sampled by the time evolution of the system. We shall compute this equilibrium ensemble, which gives a coarse-grained approximation to the detailed time-dependent dynamics of the system, by the methods of statistical mechanics. Specifically, we shall assume that in the equilibrium ensemble, the a priori distribution of matrix variables is uniform over the matrix operator phase space, so that the equilibrium ensemble is determined solely by maximizing the combinatoric probability subject to the constraints imposed by the generic conservation laws. Liouville's theorem implies that if the assumption of a uniform a priori probability distribution is made at one time, then it is valid at all later times, assuring the consistency of the concept of an equilibrium ensemble. We do not propose to address the question of how the uniform sampling of phase space configurations by the system dynamics (the so-called ergodic hypothesis) comes about, since justifying this is still an actively studied issue in rigorous statistical mechanics.

More specifically, let $d\mu = d\mu[\{x_r\}]$ denote the operator phase space measure discussed in detail in the preceding section. In what follows we shall not need the specific form of this measure, but only the properties that it obeys Liouville's theorem, and that the measure is invariant under infinitesimal matrix operator shifts δx_r, that is

$$d\mu[\{x_r + \delta x_r\}] = d\mu[\{x_r\}]. \tag{4.8}$$

(This property will be used later on, when we discuss the equipartition or Ward identities.) For a system in statistical equilibrium, there is an equilibrium phase space density distribution $\rho[\{x_r\}] \geq 0$, such that

$$dP = d\mu[\{x_r\}]\rho[\{x_r\}] \tag{4.9a}$$

is the infinitesimal probability of finding the system in the operator phase space volume element $d\mu$, with the total probability equal to unity

$$1 = \int dP = \int d\mu[\{x_r\}]\rho[\{x_r\}]. \tag{4.9b}$$

The first task in a statistical mechanical analysis is to determine the equilibrium distribution ρ.

Since equilibrium implies that $\dot{\rho} = 0$, the equilibrium distribution can depend only on conserved operators and conserved trace functionals. In the generic case for a matrix model that is global unitary invariant, we have seen in Chapters 1 and 2 that, in addition to the conserved trace Hamiltonian $\mathbf{H}$, there is a conserved operator $\tilde{C}$, which is anti-self-adjoint when fermionic adjointness is assigned as in Eqs. (2.4b) and (2.18d). If the model is assumed to be constructed in a way that balances the numbers of fermionic qs and ps, there is additionally a conserved trace "fermion number" $\mathbf{N}$. When the discrete mode index r labels infinitesimal boxes in a spatial manifold, and the model is Poincaré invariant on this manifold, there will also be a locally conserved trace energy-momentum tensor density $T^{\mu\sigma}$, from which one can obtain by spatial integration not only the conserved trace Hamiltonian, but also conserved trace generators for three-momentum $\vec{\mathbf{P}}$, total angular momentum $\vec{\mathbf{J}}$, and Lorentz boosts $\vec{\mathbf{K}}$.

We shall assume henceforth a statistical ensemble that is at rest. That is, we take an ensemble that is neither spatially translating, rotating, nor accelerating, and so the ensemble averages of $\vec{\mathbf{P}}$, $\vec{\mathbf{J}}$, and $\vec{\mathbf{K}}$ are zero. When $\mathbf{H}$ contains no preferred spatial direction, this is achieved by taking a distribution function that has no dependence on $\vec{\mathbf{P}}$, $\vec{\mathbf{J}}$, and $\vec{\mathbf{K}}$. Thus, it suffices to consider distribution functions that depend only on the generic conserved quantities $\mathbf{H}$, $\mathbf{N}$, and $\tilde{C}$, and so the general equilibrium distribution has the form

$$\rho = \rho(\tilde{C}, \mathbf{H}, \mathbf{N}). \tag{4.9c}$$

Note that for a system at rest, with $\vec{\mathbf{P}} = \mathbf{0}$, the Lorentz invariant trace mass $[\mathbf{P}^{0^2} - \vec{\mathbf{P}}^2]^{\frac{1}{2}}$ reduces to $\mathbf{H}$, and does not appear as an additional argument of the distribution function.

Since the ensemble picks out a preferred frame, which we tentatively identify with the frame in which the cosmological black-body radiation is isotropic, it is clearly not Lorentz invariant, even when (as we shall always assume) the underlying trace dynamics action and equations of motion are Lorentz invariant. We shall argue later on that this Lorentz noninvariance coming from the $\mathbf{H}$ dependence of the ensemble decouples from the emergent quantum theory, the structure of which is governed by the $\tilde{C}$ dependence of the ensemble. Since $\tilde{C}$ is Lorentz invariant, by virtue of its invariance under global unitary invariant canonical transformations that generate Lorentz transformations of the trace action, the emergent quantum theory will also be Lorentz invariant.

In addition to its dependence on the dynamical variables, ρ can also depend on constant parameter values, with the functional form of ρ and the values of the parameters together defining the statistical ensemble. Including a traceless anti-self-adjoint operator parameter $\tilde{\lambda}$ and real number parameters τ and η, which correspond to the respective structures of $\tilde{C}$, **H**, and **N**, the general form of the equilibrium ensemble corresponding to Eq. (4.9c) is

$$\rho = \rho(\tilde{C}, \tilde{\lambda}; \mathbf{H}, \tau; \mathbf{N}, \eta). \tag{4.9d}$$

The parameter τ has the dimensions of inverse mass, and is a trace dynamics analog of the temperature in statistical mechanics, while the parameter η is dimensionless, and is a trace dynamics analog of the chemical potential in statistical mechanics. The operator parameter $\tilde{\lambda}$ has the dimensions of inverse action, and is a feature of trace dynamics that has no analog in standard classical statistical mechanics. As we noted a little earlier in our discussion, it is the $\tilde{C}$ dependence of the canonical ensemble that gives rise to emergent quantum mechanics, and we shall find that it is the fact that $\tilde{C}$ has the dimensions of action that gives rise to the appearance of Planck's constant in this quantum mechanics. In the canonical ensemble, we shall show shortly that the dependence on $\tilde{C}$ and $\tilde{\lambda}$ is only through the single real number $\mathrm{Tr}\tilde{\lambda}\tilde{C}$, and so specializing to this case, Eq. (4.9d) becomes

$$\rho = \rho(\mathrm{Tr}\tilde{\lambda}\tilde{C}; \mathbf{H}, \tau; \mathbf{N}, \eta). \tag{4.9e}$$

(When $\tilde{C}$ has a self-adjoint part, the appropriate generalization of Eq. (4.9e) is obtained by dividing $\tilde{C}$ into self-adjoint and anti-self-adjoint parts $\tilde{C}^{\mathrm{sa}}$ and $\tilde{C}^{\mathrm{asa}}$, and including two operator parameters $\tilde{\lambda}^{\mathrm{sa}}$ and $\tilde{\lambda}^{\mathrm{asa}}$ with corresponding adjointness properties. This gives as the appropriately extended form of the canonical ensemble

$$\rho = \rho(\mathrm{Tr}\tilde{\lambda}^{\mathrm{sa}}\tilde{C}^{\mathrm{sa}}; \mathrm{Tr}\tilde{\lambda}^{\mathrm{asa}}\tilde{C}^{\mathrm{asa}}; \mathbf{H}, \tau; \mathbf{N}, \eta). \tag{4.9f}$$

We shall assume that a possible self-adjoint contribution $\mathrm{Tr}\tilde{\lambda}^{\mathrm{sa}}\tilde{C}^{\mathrm{sa}}$ to ρ is sufficiently small that it can be ignored in our present discussion; a role for a very small self-adjoint term is discussed later on in Chapter 6.)

We shall now show that some significant consequences follow from the general form of Eq. (4.9e), together with the assumption that **H** is constructed from the operators $\{x_r\}$ using only c-number coefficients (as needed to insure its global unitary invariance). For a general operator $\mathcal{O}$, let us define the ensemble average

$\langle \mathcal{O} \rangle_{\text{AV}}$ by

$$\langle \mathcal{O} \rangle_{\text{AV}} = \int d\mu \rho \mathcal{O}. \tag{4.10a}$$

Then when $\mathcal{O}$ is constructed from the $\{x_r\}$ using only c-number coefficients, since the only fixed operator present in the integrands on the right-hand side of Eq. (4.10a) is $\tilde{\lambda}$, the ensemble average $\langle \mathcal{O} \rangle_{\text{AV}}$ will have the form

$$\langle \mathcal{O} \rangle_{\text{AV}} = F_{\mathcal{O}}(\tilde{\lambda}), \tag{4.10b}$$

with the function $F_{\mathcal{O}}$ constructed from its argument using only c-number coefficients (in which we include the τ and η dependence). This can be simply proved by making a global unitary transformation U on all of the phase space integration variables in the numerator of Eq. (4.10a), which when $\mathcal{O}$ is constructed using only c-number coefficients, and using the unitary invariance of the integration measure $d\mu$, implies that $F_{\mathcal{O}}(\tilde{\lambda}) = U F_{\mathcal{O}}(U^{-1}\tilde{\lambda}U)U^{-1}$, which shows that the matrix structure of $F_{\mathcal{O}}(\tilde{\lambda})$ can only involve $\tilde{\lambda}$ itself. As a consequence, the ensemble parameter $\tilde{\lambda}$ commutes with $\langle \mathcal{O} \rangle_{\text{AV}}$

$$[\tilde{\lambda}, \langle \mathcal{O} \rangle_{\text{AV}}] = 0. \tag{4.10c}$$

Let us now exploit the fact that the anti-self-adjoint operator $\tilde{\lambda}$ can always be diagonalized by a unitary transformation on the basis of states in Hilbert space, which we have seen is also an invariance of the integration measure $d\mu$. Specializing to $\mathcal{O} = \tilde{C}$, the functional relationship of Eq. (4.10b) between $\tilde{\lambda}$ and $\langle \tilde{C} \rangle_{\text{AV}}$ then implies that $\langle \tilde{C} \rangle_{\text{AV}}$ is diagonal in this basis as well. This brings $\langle \tilde{C} \rangle_{\text{AV}}$ into the following canonical polar form, written in terms of a real diagonal and non-negative "magnitude" operator D_{eff} and a unitary diagonal "phase" operator i_{eff}

$$\langle \tilde{C} \rangle_{\text{AV}} = i_{\text{eff}} D_{\text{eff}}, \quad \text{Tr}(i_{\text{eff}} D_{\text{eff}}) = 0,$$
$$i_{\text{eff}} = -i_{\text{eff}}^{\dagger}, \quad i_{\text{eff}}^2 = -1, \quad [i_{\text{eff}}, D_{\text{eff}}] = 0. \tag{4.11a}$$

(Without first diagonalizing, this representation also follows from the fact that a general anti-self-adjoint matrix A can be written as $A = UD$, where $D = (A^{\dagger}A)^{\frac{1}{2}} = (-A^2)^{\frac{1}{2}}$ is non-negative, and where $U = AD^{-1}$ is unitary with $U^2 = -1$.) Although the case of general D_{eff}, which corresponds to an ensemble that is asymmetrical in the Hilbert space basis, is interesting, we shall restrict ourselves henceforth to the special case in which D_{eff} is a real constant times the unit operator. In other words, *we assume that the ensemble does not favor any state in Hilbert space over any other*, as a result of initial conditions for the underlying dynamics. (Presumably these initial conditions arise at the origin of the universe in the "Big Bang," a better understanding of which could ultimately lead to a justification of our structural assumption that D_{eff} is described by a single real number constant.)

Since we shall see in Chapter 5 that this real constant, which has the dimensions of action, plays the role of Planck's constant in the emergent quantum mechanics derived from the canonical ensemble, we shall denote it by $\hbar$, and so we have

$$\langle \tilde{C} \rangle_{AV} = i_{\text{eff}} \hbar ,$$
$$\text{Tr} i_{\text{eff}} = 0. \tag{4.11b}$$

Since the relations $i_{\text{eff}} = -i_{\text{eff}}^{\dagger}$ and $i_{\text{eff}}^2 = -1$ imply that i_{eff} can be diagonalized to take the form $i\,\text{diag}(\pm 1, \pm 1, \ldots, \pm 1)$, the condition $\text{Tr} i_{\text{eff}} = 0$ of Eq. (4.11b) requires that the plus and minus eigenvalues must be paired so as to give a vanishing trace. Therefore the dimension N of the underlying matrix Hilbert space must be even, say $N = 2K$, and i_{eff} diagonalizes to the form

$$i_{\text{eff}} = i\,\text{diag}(1, -1, 1, -1, \ldots, 1, -1), \tag{4.11c}$$

with equal numbers of eigenvalues 1 and -1 along the principal diagonal. The restriction to even N is a direct result of our assumption that the magnitude matrix D_{eff} in Eq. (4.11a) is a multiple of the unit matrix; if one were to start off with a matrix space with N odd, then $\text{Tr}(i_{\text{eff}} D_{\text{eff}}) = 0$ from Eq. (4.11a) would require D_{eff} to have one null eigenvalue, since a one-dimensional traceless matrix must vanish. As we shall discuss in detail in Section 4.5, the assumption that D_{eff} is a multiple of the unit matrix is equivalent to the assumption that the canonical ensemble breaks the global unitary invariance group $U(N)$ to the maximal subgroup that is consistent with having a non-vanishing value of $\langle \tilde{C} \rangle_{AV}$.

We turn now to the calculation of the functional form of ρ in the canonical ensemble, which is the ensemble relevant for describing the behavior of a large system that is a subsystem of a very much larger system. The form of ρ is determined by maximizing the entropy (for standard statistical mechanical discussions, see Sommerfeld, 1956 and ter Haar, 1995)

$$S = -\int d\mu \rho \log \rho, \tag{4.12a}$$

subject to the constraints

$$\int d\mu \rho = 1,$$
$$\int d\mu \rho \tilde{C} = \langle \tilde{C} \rangle_{AV},$$
$$\int d\mu \rho \mathbf{H} = \langle \mathbf{H} \rangle_{AV}, \tag{4.12b}$$
$$\int d\mu \rho \mathbf{N} = \langle \mathbf{N} \rangle_{AV}.$$

The standard procedure is to impose the constraints with Lagrange multipliers $\theta, \tilde{\lambda}, \tau, \eta$ by writing

$$\mathcal{F} = \int d\mu \rho \log \rho + \theta \int d\mu \rho + \int d\mu \rho \mathrm{Tr}\tilde{\lambda}\tilde{C} + \tau \int d\mu \rho \mathbf{H} + \eta \int d\mu \rho \mathbf{N},$$

(4.13a)

and maximizing $-\mathcal{F}$ (or equivalently, minimizing $\mathcal{F}$), treating all variations of ρ as independent. Equating to zero the derivative of Eq. (4.13a) with respect to ρ then implies

$$\rho = \exp(-1 - \theta - \mathrm{Tr}\tilde{\lambda}\tilde{C} - \tau\mathbf{H} - \eta\mathbf{N}),$$

(4.13b)

which on imposing the condition that ρ be normalized to unity gives finally

$$\rho = Z^{-1}\exp(-\mathrm{Tr}\tilde{\lambda}\tilde{C} - \tau\mathbf{H} - \eta\mathbf{N}),$$
$$Z = \int d\mu \exp(-\mathrm{Tr}\tilde{\lambda}\tilde{C} - \tau\mathbf{H} - \eta\mathbf{N}).$$

(4.13c)

When $\tilde{C}$ has a self-adjoint part, $\mathrm{Tr}\tilde{\lambda}\tilde{C}$ in Eq. (4.13c), and in subsequent equations involving the canonical ensemble, is modified according to

$$\mathrm{Tr}\tilde{\lambda}\tilde{C} \to \mathrm{Tr}\tilde{\lambda}^{\mathrm{sa}}\tilde{C}^{\mathrm{sa}} + \mathrm{Tr}\tilde{\lambda}^{\mathrm{asa}}\tilde{C}^{\mathrm{asa}},$$

(4.13d)

with the superscripts "sa" and "asa" denoting self-adjoint and anti-self-adjoint parts respectively. As noted following Eq. (4.9f), we shall assume, until we get to Chapter 6, that any self-adjoint part of $\tilde{C}$ is very small, and shall not explicitly carry along the extra terms in our formulas.

In order for the partition function of Eq. (4.13c) to be well defined, the trace Hamiltonian must be bounded from below in the bosonic variables, and must increase rapidly enough as these variables become infinite for the bosonic integrals to converge at $\pm\infty$. Since $\tilde{C}$ grows as the bilinear product qp for each bosonic canonical pair q, p, as long as $\mathbf{H}$ contains terms at least quadratic in p and q, there will be an interval of convergence in the parameter $\tilde{\lambda}$. Then, by choosing an appropriate $\tilde{\lambda}$ within this interval of convergence, we in general expect that the ensemble average of $\tilde{C}$ can be made to take any specified value, and so the condition on this average given in the second equation of Eq. (4.12b) can be satisfied. Requiring that ρ should be non-negative may also place restrictions on the fermionic structure of $\mathbf{H}$, since we must require, for example, that after the fermionic integrations in the partition function have been carried out, the effective measure thus defined on the bosonic variables should be non-negative. This can be guaranteed, for example, if the fermionic structure has a reflection symmetry which gives the fermionic integral the form of a real number squared, in analogy with the behavior of the fermionic Euclidean functional integral for vector-like quantum field theories studied in Vafa and Witten (1983, 1984). For example, if the trace

Hamiltonian has the bilinear form in the fermionic variables given in Eq. (2.1a), then the entire exponent in the partition function Z of Eq. (4.13c) is a quadratic form in the fermionic variables x_r, permitting the fermionic integral with the measure of Eq. (4.7d) to be evaluated as a determinant. Positivity is then guaranteed if there is a reflection symmetry that leads to the factorization of this determinant into the absolute value squared of a smaller determinant.

From Eq. (4.13c) we can derive some elementary statistical properties of the equilibrium ensemble. For the entropy S, we find

$$S = -\langle \log \rho \rangle_{AV} = \log Z + \mathrm{Tr}\tilde{\lambda}\langle \tilde{C} \rangle_{AV} + \tau \langle \mathbf{H} \rangle_{AV} + \eta \langle \mathbf{N} \rangle_{AV}. \qquad (4.14a)$$

Since the ensemble averages which appear in Eq. (4.14a) are given by

$$\langle \tilde{C} \rangle_{AV} = -\frac{\delta \log Z}{\delta \tilde{\lambda}},$$

$$\langle \mathbf{H} \rangle_{AV} = -\frac{\partial \log Z}{\partial \tau},$$

$$\langle \mathbf{N} \rangle_{AV} = -\frac{\partial \log Z}{\partial \eta}, \qquad (4.14b)$$

Eq. (4.14a) takes the form

$$S = \log Z - \mathrm{Tr}\tilde{\lambda}\frac{\delta \log Z}{\delta \tilde{\lambda}} - \tau \frac{\partial \log Z}{\partial \tau} - \eta \frac{\partial \log Z}{\partial \eta}. \qquad (4.14c)$$

Thus the entropy is a thermodynamic quantity determined solely by the partition function. Taking second derivatives of the partition function, we can similarly derive the thermodynamic formulas for the averaged mean square fluctuations of the conserved quantities $\tilde{C}$, $\mathbf{H}$, and $\mathbf{N}$

$$\Delta^2_{\mathrm{Tr}\tilde{P}\tilde{C}} \equiv \langle (\mathrm{Tr}\tilde{P}\tilde{C} - \langle \mathrm{Tr}\tilde{P}\tilde{C} \rangle_{AV})^2 \rangle_{AV}$$

$$= \langle (\mathrm{Tr}\tilde{P}\tilde{C})^2 \rangle_{AV} - \langle \mathrm{Tr}\tilde{P}\tilde{C} \rangle^2_{AV} = \left(\mathrm{Tr}\tilde{P}\frac{\delta}{\delta\tilde{\lambda}} \right)^2 \log Z,$$

$$\Delta^2_{\mathbf{H}} \equiv \langle (\mathbf{H} - \langle \mathbf{H} \rangle_{AV})^2 \rangle_{AV} = \langle \mathbf{H}^2 \rangle_{AV} - \langle \mathbf{H} \rangle^2_{AV} = \frac{\partial^2 \log Z}{(\partial \tau)^2}, \qquad (4.14d)$$

$$\Delta^2_{\mathbf{N}} \equiv \langle (\mathbf{N} - \langle \mathbf{N} \rangle_{AV})^2 \rangle_{AV} = \langle \mathbf{N}^2 \rangle_{AV} - \langle \mathbf{N} \rangle^2_{AV} = \frac{\partial^2 \log Z}{(\partial \eta)^2},$$

with $\tilde{P}$ an arbitrary fixed anti-self-adjoint operator. Similar expressions hold for the cross-correlations of $\mathrm{Tr}\tilde{P}\tilde{C}$, $\mathbf{H}$, and $\mathbf{N}$, for example

$$\langle (\mathrm{Tr}\tilde{P}\tilde{C} - \langle \mathrm{Tr}\tilde{P}\tilde{C} \rangle_{AV})(\mathbf{H} - \langle \mathbf{H} \rangle_{AV}) \rangle_{AV} = \langle \mathrm{Tr}\tilde{P}\tilde{C}\mathbf{H} \rangle_{AV} - \langle \mathrm{Tr}\tilde{P}\tilde{C} \rangle_{AV}\langle \mathbf{H} \rangle_{AV}$$

$$= \mathrm{Tr}\tilde{P}\frac{\delta}{\delta\tilde{\lambda}}\frac{\partial}{\tau} \log Z. \qquad (4.14e)$$

The complete set of such relations, in a more compact notation, is given in the next section. As a final remark, we note that the integration over phase space in the canonical ensemble is performed with all variables taken at a fixed time, which we choose by convention to be $t = 0$. However, by using Eq. (1.19) to formally integrate forward or backward in time, we can express any variable $x_r(t)$ at a general time t in terms of its $t = 0$ value, and thus can form averages of phase space variables at general times over the canonical ensemble.

In subsequent sections we shall follow the conventional practice of introducing for each matrix variable x_r a matrix source j_r, of the same bosonic or fermionic type and with the same adjointness properties as x_r, which can be varied and then set to zero after all variations have been performed. With the sources included, the equilibrium distribution and partition function take the form

$$\rho_j = Z_j^{-1} \exp\left(- \mathrm{Tr}\tilde{\lambda}\tilde{C} - \tau \mathbf{H} - \eta \mathbf{N} - \sum_r \mathrm{Tr}\, j_r x_r \right),$$

$$Z_j = \int d\mu \exp\left(- \mathrm{Tr}\tilde{\lambda}\tilde{C} - \tau \mathbf{H} - \eta \mathbf{N} - \sum_r \mathrm{Tr}\, j_r x_r \right). \qquad (4.15a)$$

Using the expression $\langle \mathcal{O} \rangle_{\mathrm{AV},j}$ to denote the average of a general operator over the equilibrium distribution of Eq. (4.15a) which includes sources, the variations of $\log Z_j$ with respect to its source arguments j_r are related to the averages of the x_r by

$$\epsilon_r \langle x_r \rangle_{\mathrm{AV},j} = - \frac{\delta \log Z_j}{\delta j_r}. \qquad (4.15b)$$

4.3 The microcanonical ensemble

In the previous section, we derived the canonical ensemble for trace dynamics by maximizing the entropy subject to the generic constraints, which were imposed in an averaged sense. In this section, we shall give, following Adler and Horwitz (1996), an alternative and more fundamental derivation of the canonical ensemble, by starting from the microcanonical ensemble, in which the constraints are imposed in a sharp sense. We shall see that the canonical ensemble then arises as the appropriate description of a large system in equilibrium with a much larger "bath," with the equilibrium conditions determining in an intrinsic manner the ensemble parameters, or generalized "temperatures" $\tilde{\lambda}$, τ, and η. Apart from using the microcanonical ensemble to derive the canonical ensemble in this section we shall not employ the microcanonical ensemble further in our subsequent analysis. The reason for our primary focus on the canonical ensemble is that the Ward identities that imply emergent quantum behavior, which are derived in Chapter 5, are properties of the canonical ensemble but not, as we shall see, of the microcanonical ensemble. Thus, if the microcanonical ensemble is taken to represent the entire universe, our subsequent analysis suggests that emergent quantum mechanics is a property only of subsystems of the universe that are large but still appreciably smaller than the universe as a whole.

It is convenient at this point to introduce a condensed notation for the exponent appearing in the canonical ensemble of Eq. (4.13c), which takes the anti-self-adjointness of $\tilde{C}$ and $\tilde{\lambda}$ into account. Writing as in Eq. (4.2a)

$$(\tilde{C})_{mn} = (\tilde{C})^0_{mn} + i(\tilde{C})^1_{mn},$$

$$(\tilde{\lambda})_{mn} = (\tilde{\lambda})^0_{mn} + i(\tilde{\lambda})^1_{mn}, \qquad (4.16a)$$

the anti-self-adjointness restrictions on $\tilde{C}$ and $\tilde{\lambda}$ take the form

$$(\tilde{C})^A_{mn} = -\epsilon^A(\tilde{C})^A_{nm}, \quad (\tilde{\lambda})^A_{mn} = -\epsilon^A(\tilde{\lambda})^A_{nm}, \qquad (4.16b)$$

with $\epsilon^0 = 1$ and $\epsilon^1 = -1$ as before. Then a simple calculation shows that

$$\mathrm{Tr}\tilde{\lambda}\tilde{C} = -\sum_n (\tilde{\lambda})^1_{nn}(\tilde{C})^1_{nn} - 2\sum_{n<m}[(\tilde{\lambda})^0_{nm}(\tilde{C})^0_{nm} + (\tilde{\lambda})^1_{nm}(\tilde{C})^1_{nm}], \qquad (4.16c)$$

with all the terms on the right-hand side independent. It is now convenient to introduce a vector notation for the exponent in Eq. (4.13c), by defining

$$\xi \equiv (\xi^1, \ldots, \xi^M)$$
$$\equiv \left(\mathbf{H}, \mathbf{N}, [(\tilde{C})^1_{nn}, n = 1, \ldots, N], [(\tilde{C})^0_{nm}, n < m = 1, \ldots, N],\right.$$
$$\left. [(\tilde{C})^1_{nm}, n < m = 1, \ldots, N]\right),$$
$$\sigma \equiv (\sigma^1, \ldots, \sigma^M)$$
$$\equiv \left(\tau, \eta, -[(\tilde{\lambda})^1_{nn}, n = 1, \ldots, N], -2[(\tilde{\lambda})^0_{nm}, n < m = 1, \ldots, N],\right.$$
$$\left. -2[(\tilde{\lambda})^1_{nm}, n < m = 1, \ldots, N]\right), \qquad (4.17a)$$

which permits us to write

$$\mathrm{Tr}\tilde{\lambda}\tilde{C} + \tau\mathbf{H} + \eta\mathbf{N} = \vec{\sigma}\cdot\vec{\xi}. \qquad (4.17b)$$

(When $\tilde{C}$ has a self-adjoint part, one extends the definition of ξ to include the non-vanishing matrix elements of both $\tilde{C}^{\mathrm{sa}}$ and $\tilde{C}^{\mathrm{asa}}$, and the definition of σ to include the non-vanishing matrix elements of both $\tilde{\lambda}^{\mathrm{sa}}$ and $\tilde{\lambda}^{\mathrm{asa}}$, in analogy with Eq. (4.13d).) In other words, $\vec{\xi}$ is the vector of all the real number generic conserved quantities, and $\vec{\sigma}$ is the vector of the corresponding canonical ensemble parameters (which are analogs of the inverse temperature parameter β and the chemical potential parameter μ of ordinary statistical mechanics). The dimensionality M of both vectors is $M = 2 + N^2$.

We can now introduce the microcanonical ensemble $\Gamma(\vec{\Xi})$, which is defined as the volume of the shell of phase space for the "universe" in which the conserved quantities take the sharp values $\vec{\Xi}$. In other words, we write

$$\Gamma(\vec{\Xi}) = \int d\mu \prod_{a=1}^{M} \delta(\Xi^a - \xi^a),\tag{4.18a}$$

which has an associated entropy

$$S(\vec{\Xi}) = \log \Gamma(\vec{\Xi}).\tag{4.18b}$$

Let us now divide the universe into a "system" s, which is still large in a statistical sense but is much smaller than the universe, and a "bath" b which is the complement of degrees of freedom in the universe not included in the system. We now assume that the vector of conserved quantities $\vec{\xi}$ is to a good approximation additively decomposable over the system and the bath, when both are very large. In other words, we assume that

$$\vec{\xi} \simeq \vec{\xi}_s + \vec{\xi}_b,\tag{4.19a}$$

with $\vec{\xi}_s$ and $\vec{\xi}_b$ the values of the conserved quantities appropriate to the system and to the bath, respectively. Taking the system and the bath to be defined simply by a partitioning of the canonical degrees of freedom q_r, p_r, additivity is automatic for $\mathbf{N}$ and $\tilde{C}$, which are additive sums over the degrees of freedom, but not for $\mathbf{H}$, which in general has couplings between the degrees of freedom. The assumption of Eq. (4.19a) is then that the contribution to $\mathbf{H}$ from terms that couple the system degrees of freedom to the bath degrees of freedom is much smaller than the contributions that involve solely the system degrees of freedom or solely the bath degrees of freedom. (This is almost certainly true, for example, for trace dynamics models built from conventional spacetime field theory Lagrangians, such as the models of Sections 2.3 and 2.4 and Chapter 3, when the system is defined as a bounded finite region surrounded by the bath; the coupling terms from system to bath then involve only the system degrees of freedom on the surface separating the system from the bath.) Letting $d\mu_s$ and $d\mu_b$ be the phase space measures for the system and the bath, so that $d\mu = d\mu_s d\mu_b$, and introducing a dummy variable of integration $\vec{\Xi}_s$, we can rewrite Eq. (4.18a) as

$$\Gamma(\vec{\Xi}) = \prod_a \int d\Xi_s^a \Gamma_b(\vec{\Xi} - \vec{\Xi}_s)\Gamma_s(\vec{\Xi}_s),\tag{4.19b}$$

with the system and bath microcanonical subensembles defined by

$$\Gamma_s(\vec{\Xi}_s) \equiv \int d\mu_s \prod_{a=1}^{M} \delta(\Xi_s^a - \xi_s^a),$$

$$\Gamma_b(\vec{\Xi} - \vec{\Xi}_s) \equiv \int d\mu_b \prod_{a=1}^{M} \delta(\Xi^a - \Xi_s^a - \xi_b^a). \tag{4.19c}$$

We now assume that the integrand in Eq. (4.19b) has a maximum that dominates the integral when the number of degrees of freedom is large. Although we give no a priori justification of this assumption, we shall later on show that it is self-consistent. The necessary condition for the integrand in Eq. (4.19b) to have an extremum at $\vec{\Xi}_s = \vec{X}_s$ is

$$\frac{\partial}{\partial \Xi_s^a} [\Gamma_b(\vec{\Xi} - \vec{\Xi}_s)\Gamma_s(\vec{\Xi}_s)]|_{\vec{X}_s} = 0, \tag{4.20a}$$

which can be rewritten as

$$\sigma^a \equiv \frac{\partial}{\partial \Xi_s^a} \log \Gamma_s(\vec{\Xi}_s)|_{\vec{X}_s} = \frac{\partial}{\partial \Xi^a} \log \Gamma_b(\vec{\Xi} - \vec{\Xi}_s)|_{\vec{X}_s}. \tag{4.20b}$$

Thus, at the assumed maximum, the logarithmic derivatives in Eq. (4.20b) define a set of equilibrium parameters $\vec{\sigma}$ common to the bath and the system. Recalling the entropy definition of Eq. (4.18b), we can rewrite the bath phase space volume at the extremum as

$$\Gamma_b(\vec{\Xi} - \vec{X}_s) = \exp\left(S_b(\vec{\Xi} - \vec{X}_s)\right) \simeq \exp\left(S_b(\vec{\Xi})\right) \exp\left(-\sum_a X_s^a \frac{\partial}{\partial \Xi^a} \log \Gamma_b(\vec{\Xi})\right), \tag{4.20c}$$

which, neglecting a small shift from $\vec{\Xi} - \vec{X}$ to $\vec{\Xi}$ in the definition of the equilibrium parameters $\vec{\sigma}$, gives us

$$\Gamma_b(\vec{\Xi} - \vec{X}_s) \simeq \exp\left(S_b(\vec{\Xi})\right) \exp(-\vec{\sigma} \cdot \vec{X}_s). \tag{4.20d}$$

By continuity, Eq. (4.20d) implies that for $\vec{\Xi}_s$ in the neighborhood of the extremum $\vec{X}_s$, we similarly have

$$\Gamma_b(\vec{\Xi} - \vec{\Xi}_s) \simeq \exp\left(S_b(\vec{\Xi})\right) \exp(-\vec{\sigma} \cdot \vec{\Xi}_s). \tag{4.20e}$$

Returning now to Eq. (4.19b) and substituting the approximate form of Eq. (4.20e) for the bath phase space volume factor, we get

$$\Gamma(\vec{\Xi}) \simeq \exp\left(S_b(\vec{\Xi})\right) Z_s, \tag{4.21a}$$

with Z_s the integral defined by

$$Z_s = \prod_a \int d\Xi_s^a \exp(-\vec{\sigma} \cdot \vec{\Xi}_s) \Gamma_s(\vec{\Xi}_s).$$

(4.21b)

On substituting Eq. (4.19c) for the system phase space volume $\Gamma_s(\vec{\Xi}_s)$ and carrying out the integration over the dummy variables Ξ_s^a, we can rewrite Z_s as

$$Z_s = \int d\mu_s \exp(-\vec{\sigma} \cdot \vec{\xi}_s).$$

(4.21c)

We conclude that when the system and bath are in equilibrium, and the overall "universe" comprising the system and bath is in a microcanonical ensemble, the system variables are weighted in the phase space integral according to the normalized distribution

$$\rho_s = Z_s^{-1} \exp(-\vec{\sigma} \cdot \vec{\xi}_s),$$

(4.22)

which defines the standard canonical ensemble. Since all of the above manipulations go through if $d\mu_s$ is replaced by $d\mu_s f_s$, with f_s any function of the system variables, we have shown that the average $\langle f_s \rangle_{AV}$ defined in the microcanonical ensemble

$$\langle f_s \rangle_{AV} \equiv \frac{\int d\mu \prod_{a=1}^{M} \delta(\Xi^a - \xi^a) f_s}{\int d\mu \prod_{a=1}^{M} \delta(\Xi^a - \xi^a) 1}$$

(4.23a)

can be equivalently calculated as

$$\langle f_s \rangle_{AV} \simeq \int d\mu_s \rho_s f_s.$$

(4.23b)

This justifies the use of the canonical ensemble in calculating thermodynamic averages of system quantities.

As a consistency check on the calculation, we must verify that within our approximations, the extremum of Eq. (4.20a) is a maximum. Using the approximated form of the integrand in Eq. (4.21b), the condition for an extremum is

$$\frac{\partial}{\partial \Xi_s^a} [-\vec{\sigma} \cdot \vec{\Xi}_s + \log \Gamma_s(\vec{\Xi}_s)] = 0.$$

(4.24a)

In other words, the extremum $\vec{\Xi}_s = \vec{X}_s$ is the solution of the equation

$$\sigma^a = \frac{\partial}{\partial \Xi_s^a} \log \Gamma_s(\vec{\Xi}_s),$$

(4.24b)

in agreement with the definition of Eq. (4.20b). In order for the extremum to be a maximum, the matrix of second derivatives

$$M_{ab}(\vec{\Xi}_s) \equiv \frac{\partial}{\partial \Xi_s^a} \frac{\partial}{\partial \Xi_s^b} [-\vec{\sigma} \cdot \vec{\Xi}_s + \log \Gamma_s(\vec{\Xi}_s)] = \frac{\partial}{\partial \Xi_s^a} \frac{\partial}{\partial \Xi_s^b} \log \Gamma_s(\vec{\Xi}_s)$$

$$(4.24c)$$

must be negative definite. Regarding $\vec{\sigma}$ now as a variable $\vec{\sigma}(\vec{\Xi}_s)$ defined by Eq. (4.24b) for $\vec{\Xi}_s$ away from the extremum $\vec{X}_s$, the matrix of second derivatives can be rewritten as

$$M_{ab}(\vec{\Xi}_s) = \frac{\partial}{\partial \Xi_s^a} \sigma^b. \qquad (4.24d)$$

This will be negative definite provided that the inverse matrix $\partial \Xi_s^a / \partial \sigma^b$ is negative definite (and bounded), with $\vec{\Xi}_s = \vec{\Xi}_s(\vec{\sigma})$ the location of the maximum of the integrand in Eq. (4.21b), which for a large system is closely approximated by the canonical ensemble average $\langle \vec{\xi}_s \rangle_{AV}$. But from Eq. (4.21c), we see that

$$\langle \xi_s^a \rangle_{AV} = -\frac{\partial \log Z_s}{\partial \sigma^a}, \qquad (4.25a)$$

and differentiating again

$$\frac{\partial \langle \xi_s^a \rangle_{AV}}{\partial \sigma^b} = -\frac{\partial^2 \log Z_s}{\partial \sigma^a \partial \sigma^b} = -\langle \xi_s^a \xi_s^b \rangle_{AV} + \langle \xi_s^a \rangle_{AV} \langle \xi_s^b \rangle_{AV}$$

$$= -\langle (\xi_s^a - \langle \xi_s^a \rangle_{AV})(\xi_s^b - \langle \xi_s^b \rangle_{AV}) \rangle_{AV}, \qquad (4.25b)$$

which is negative definite and bounded. Thus the assumption that the extremum in the phase space integral is a maximum is self-consistent. Referring back to the correlation formulas of Eqs. (4.14d,e), we see that Eq. (4.25b) gives the most general such formula in our condensed notation.

4.4 Gauge fixing in the partition function*

Up to this point, in discussing the statistical mechanics of trace dynamics we have assumed that one is dealing with an unconstrained system, leading to the generic form of the canonical ensemble given in Eq. (4.13c). In order to apply Eq. (4.13c) directly to a constrained system, one must first explicitly eliminate the constraints. A simple example where this is possible is provided by the trace dynamics transcription of supersymmetric Yang–Mills theory, discussed in Section 3.2. (Further examples of gauge invariant trace dynamics models are given in Appendix E.) In axial gauge, where $A_3 = 0$, the covariant derivative D_3 simplifies to $D_3 = \partial_3$, allowing the constraint of Eq. (3.10b) to be eliminated, giving the explicit expression for the trace Hamiltonian of Eqs. (3.11b,c), and the corresponding expression for the conserved operator $\tilde{C}$ of Eq. (3.13a). The axial gauge partition function is then given by Eq. (4.13c), with the bosonic phase space integration measure $d\mu_B$ given by

$$d\mu_{B\,\text{axial}} = \prod_{\vec{x}} \prod_{\ell=1}^{2} dA_\ell(\vec{x}) dp_{A_\ell(\vec{x})}. \tag{4.26}$$

The problem addressed in this section is how to generalize the axial gauge partition function to other gauges in which it may not be possible to explicitly integrate out the constraint (Adler, 1998). The problem of correctly incorporating a gauge invariance group with a continuous infinity of group parameters and an infinite invariant group volume is a familiar one in the theory of path integrals, and we shall use methods similar to the ones employed there to give a solution. However, since the partition function singles out a Lorentz frame, we will have to make a restriction not encountered in the Lorentz scalar path integral case, namely we will consider only nontemporal gauge conditions that do not involve the scalar potential A_0. This still allows us to consider gauge transformations that rotate the axial gauge axis, or that transform to rotationally invariant gauges such as Coulomb gauge. We shall also make the further assumption that the allowed gauge transformations leave invariant the surface integral which, according to Eq. (3.13b), determines $\tilde{C}$, thus placing a restriction on the gauge transformation at the point at infinity. We proceed by developing an analog of the standard De Witt–Faddeev–Popov method to write the axial gauge partition function in a general nontemporal gauge, subject to the surface term restriction just stated. Since we have seen in Chapter 3 that trace dynamics incorporates rigid supersymmetry, and since Becchi–Rouet–Stora–Tyutin (BRST) invariance is a particular rigid supersymmetry transformation, we shall find that the generalized expression for the partition function, when reexpressed in terms of ghost fermions, admits a BRST invariance. As in our preceding discussion, we assume convergence of the partition function, which may well require restrictions on the class of trace Hamiltonians being considered.

To express the partition function in a general nontemporal gauge, we follow closely the treatment of the De Witt–Faddeev–Popov construction in the familiar functional integral case, as given in the text of Weinberg (1996). Let us consider the integral

$$Z_G = \int d\mu_B d\mu_F \mathcal{K}[f(A_\ell)]\delta(Y) \det \mathcal{F}[A_\ell] \exp(-\text{Tr}\tilde{\lambda}\tilde{C} - \tau\mathbf{H} - \eta\mathbf{N}),$$

$$d\mu_B = \prod_{\vec{x}} \prod_{\ell=1}^{3} dA_\ell(\vec{x}) dp_{A_\ell(\vec{x})}, \quad d\mu_F = \prod_{\vec{x}} \prod_{d=1}^{4} d\chi^d(\vec{x}), \tag{4.27a}$$

with $\tilde{C}$ given by Eq. (3.13b) and with the constraint Y given by

$$Y \equiv \sum_{\ell=1}^{3} D_\ell p_{A_\ell} + 2\bar{\chi}\gamma^0\chi. \tag{4.27b}$$

The trace Hamiltonian $\mathbf{H}$ in Eq. (4.27a) is given by Eq. (3.11b), with the gauge part $\mathbf{H}_A$ given by

$$\mathbf{H}_A = \int d^3x \, \mathrm{Tr}\left(-\frac{g^2}{2}\sum_{\ell=1}^{3} p_{A_\ell}^2 - \frac{1}{4g^2}\sum_{\ell,m=1}^{3} F_{\ell m}^2\right), \tag{4.27c}$$

which is valid in a general gauge on the constraint surface $Y = 0$ selected by the delta function in Eq. (4.27a). The delta function of the anti-self-adjoint matrix valued argument Y appearing in Eq. (4.27a) is given, in terms of ordinary delta functions of the real (R) and imaginary (I) parts of the matrix elements, by

$$\delta(Y) = \prod_{m<n} \delta\big((Y_R)_{mn}\big) \prod_{m\leq n} \delta\big((Y_I)_{mn}\big). \tag{4.28a}$$

The integration measures over the anti-self-adjoint matrix A_ℓ and the self-adjoint matrix χ^d are defined by

$$dA_\ell = \prod_{m<n} d(A_{\ell R})_{mn} \prod_{m\leq n} d(A_{\ell I})_{mn}, \quad d\chi^d = \prod_{m<n} d(\chi_I^d)_{mn} \prod_{m\leq n} d(\chi_R^d)_{mn}, \tag{4.28b}$$

with the integration measure for dp_{A_ℓ} similar in structure to dA_ℓ. The function $\mathcal{K}[f]$ is an arbitrary integrable scalar valued function of the matrix valued argument $f(A_\ell)$, which is used to specify the gauge condition. We shall treat f as a column vector f_α with α a composite index formed from the matrix row and column indices m, n; the argument $\mathcal{F}[A_\ell]$ of the De Witt–Faddeev–Popov determinant $\det \mathcal{F}[A_\ell]$ appearing in Eq. (4.27a) is then defined in terms of f by the expression

$$\mathcal{F}_{\alpha\vec{x},\beta\vec{y}}[A_\ell] \equiv \frac{\delta f_\alpha\big(A_\ell(\vec{x}) + D_\ell\Lambda(\vec{x})\big)}{\delta\Lambda_\beta(\vec{y})}\bigg|_{\Lambda=0}, \tag{4.28c}$$

where δ is the usual functional derivative and β is the composite of the row and column indices of the infinitesimal gauge transformation matrix Λ.

We now demonstrate two properties of the integral Z_G defined in Eq. (4.27a): (i) first, we show that when the gauge fixing functions $\mathcal{K}[f]$ and $f(A_\ell)$ are chosen to correspond to the axial gauge condition, then Eq. (4.27a) reduces (up to an overall constant) to the axial gauge partition function; (ii) second, we show that Z_G is in fact independent of the function $f(A_\ell)$, and depends on the function $\mathcal{K}[f]$ only through an overall constant. These two properties together imply that

Z_G gives the wanted extension of the axial gauge partition function to general nontemporal gauges.

To establish property (i), we make the conventional axial gauge choice

$$\mathcal{K}[f(A_\ell)] = \delta(A_3) = \prod_{m<n} \delta((A_{3R})_{mn}) \prod_{m\leq n} \delta((A_{3I})_{mn}), \qquad (4.29a)$$

so that

$$\int dA_3 \mathcal{K}[f(A_\ell)] = \int dA_3 \delta(A_3) = 1. \qquad (4.29b)$$

With this gauge choice

$$D_3 p_{A_3} = \partial_3 p_{A_3}, \qquad (4.29c)$$

which implies that

$$\delta(Y) = \delta\left(\partial_3 p_{A_3} + \sum_{\ell=1}^{2} D_\ell p_{A_\ell} + 2\bar{\chi}\gamma^0\chi\right)$$

$$= |\partial_3|^{-1}\delta\left(p_{A_3} + \frac{1}{2}\int dz'\epsilon(z-z')\left(\sum_{\ell=1}^{2} D_\ell p_{A_\ell} + 2\bar{\chi}\gamma^0\chi\right)\right). \qquad (4.29d)$$

Hence the integral over p_{A_3} in Z_G can be done explicitly, giving (up to an overall constant factor coming from the Jacobian $|\partial_3|^{-1}$) the expression

$$Z_G = \int d\mu_B \,_{\text{axial}} d\mu_F$$

$$\times \exp(-\text{Tr}\tilde{\lambda}\tilde{C} - \tau\mathbf{H} - \eta\mathbf{N})|_{A_3=0;\ p_{A_3}=-\frac{1}{2}\int dz'\epsilon(z-z')\left(\sum_{\ell=1}^{2} D_\ell p_{A_\ell}+2\bar{\chi}\gamma^0\chi\right)}, \qquad (4.30)$$

which agrees (recalling from Eq. (3.3) that $2\bar{\chi}\gamma^0\chi = -(p_\chi\chi + \chi^T p_\chi^T)$) with the axial gauge partition function constructed from $\mathbf{H}_A$ of Eq. (3.11c).

To establish property (ii), we first examine the gauge transformation properties of the various factors in the integral defining Z_G. We begin with the integration measure $d\mu_B$. Under the infinitesimal gauge transformation (with Λ anti-self-adjoint)

$$A_\ell \to A_\ell + D_\ell\Lambda = A_\ell + \partial_\ell\Lambda + [A_\ell, \Lambda], \qquad (4.31a)$$

the inhomogeneous term $\partial_\ell\Lambda$ does not contribute to the transformation of the differential dA_ℓ. Therefore dA_ℓ obeys the homogeneous transformation law $dA_\ell \to dA_\ell + \Delta_\ell$, with

$$\Delta_\ell \equiv [dA_\ell, \Lambda]. \qquad (4.31b)$$

Hence to first order in Λ, the Jacobian of the transformation (calculated by the same reasoning that led from Eq. (4.4) to Eq. (4.5a)) is

$$J = 1 + \sum_{m<n} \frac{\partial(\Delta_{\ell R})_{mn}}{\partial(dA_{\ell R})_{mn}} + \sum_{m \leq n} \frac{\partial(\Delta_{\ell I})_{mn}}{\partial(dA_{\ell I})_{mn}}$$

$$= 1 + \left(\sum_{m<n} + \sum_{m \leq n} \right) [(\Lambda_R)_{nn} - (\Lambda_R)_{mm}]$$

$$= 1, \tag{4.31c}$$

since the anti-self-adjointness of Λ implies that $(\Lambda_R)_{nm} = -(\Lambda_R)_{mn}$, and so the diagonal matrix elements $(\Lambda_R)_{nn}$ are all zero. Thus each factor $dA_\ell(\vec{x})$ in the integration measure is gauge invariant. A similar argument applies to each factor $dp_{A_\ell(\vec{x})}$ in the bosonic integration measure, and also to the fermionic integration measure $d\mu_F$ and to the factor $\delta(Y)$ in the integrand, since χ and Y obey the homogeneous gauge transformation laws $\chi \to \chi + [\chi, \Lambda]$, $Y \to Y + [Y, \Lambda]$. Turning to the exponential, the terms $\mathrm{Tr}\, p_{A_\ell}^2$, $\mathrm{Tr}\, F_{\ell m}^2$, and $\mathrm{Tr}\chi^T \chi$ are gauge invariant, and so the trace Hamiltonian **H** and the trace fermion number **N** are gauge invariant. By hypothesis, the surface term determining $\tilde{C}$ is left invariant by the class of gauge transformations under consideration. To summarize, we see that the integral Z_G has the form

$$Z_G = \int d\mu \mathcal{G}[A_\ell] \mathcal{K}[f(A_\ell)] \det \mathcal{F}[A_\ell] \tag{4.32a}$$

with the integration measure $d\mu = d\mu_B d\mu_F$ and the integrand factor

$$\mathcal{G}[A_\ell] = \delta(Y) \exp(-\mathrm{Tr}\tilde{\lambda}\tilde{C} - \tau\mathbf{H} - \eta\mathbf{N}) \tag{4.32b}$$

both gauge invariant. Hence Z_G has exactly the form assumed in the discussion of Weinberg (1996), and the proof given there completes the demonstration of property (ii).

Continuing to follow the standard path integral analysis, let us represent the De Witt–Faddeev–Popov determinant $\det \mathcal{F}[A_\ell]$ as an integral over complex anti-self-adjoint fermionic ghost matrices ω and $\tilde{\omega}$, by writing

$$\det \mathcal{F}[A_\ell] = \int d\omega d\tilde{\omega} \exp\left(\int d^3x d^3y \tilde{\omega}_\alpha(\vec{x}) \mathcal{F}_{\alpha \vec{x}, \beta \vec{y}}[A_\ell] \omega_\beta(\vec{y}) \right), \tag{4.33a}$$

with

$$d\omega = \prod_{\vec{x}} \prod_{m,n} d\omega_{mn}(\vec{x}) \propto \prod_{\vec{x}} \left(\prod_{m<n} d\,\mathrm{Re}\omega_{mn}(\vec{x}) d\mathrm{Im}\omega_{mn}(\vec{x}) \right) \left(\prod_n d\mathrm{Im}\omega_{nn}(\vec{x}) \right),$$

$$d\tilde{\omega} = \prod_{\vec{x}} \prod_{m,n} d\tilde{\omega}_{mn}(\vec{x}) \propto \prod_{\vec{x}} \left(\prod_{m<n} d\,\mathrm{Re}\tilde{\omega}_{mn}(\vec{x}) d\mathrm{Im}\tilde{\omega}_{mn}(\vec{x}) \right) \left(\prod_n d\mathrm{Im}\tilde{\omega}_{nn}(\vec{x}) \right).$$

$$\tag{4.33b}$$

Let us also take for $\mathcal{K}[f]$ the usual Gaussian

$$\mathcal{K}[f] = \exp\left[-\frac{1}{2\xi}\int d^3x \operatorname{Tr} f\left(A_\ell(\vec{x})\right)^2\right], \qquad (4.33c)$$

and for $f(A_\ell)$ the linear gauge condition

$$f(A_\ell) = \sum_\ell L^\ell A_\ell, \qquad (4.33d)$$

in which L^ℓ can be either a fixed vector (such as $\delta_{\ell 3}$ in axial gauge) or a differential operator (such as ∂_ℓ in Coulomb gauge), and a summation of ℓ from 1 to 3 is understood. With this choice of $f(A_\ell)$, we find from Eq. (4.28c) that

$$\mathcal{F}_{nm\vec{x},pq\vec{y}}[A_\ell] = \frac{\delta f_{nm}\left(A_\ell(\vec{x}) + D_\ell \Lambda(\vec{x})\right)}{\delta \Lambda_{pq}(\vec{y})}$$

$$= \sum_\ell L^\ell_{\vec{x}}\left(\frac{\partial \delta(\vec{x}-\vec{y})}{\partial x^\ell}\delta_{np}\delta_{mq} + \delta(\vec{x}-\vec{y})[(A_\ell)_{np}\delta_{mq} - \delta_{np}(A_\ell)_{qm}]\right),$$

$$(4.34a)$$

which when substituted into the exponent in Eq. (4.33a) gives, after integration by parts

$$\int d^3x\, d^3y\, \tilde{\omega}_\alpha(\vec{x})\mathcal{F}_{\alpha\vec{x},\beta\vec{y}}[A_\ell]\omega_\beta(\vec{y}) = \int d^3x \operatorname{Tr}\overline{\omega}(\vec{x})\sum_\ell L^\ell D_\ell \omega(\vec{x}), \qquad (4.34b)$$

where we have defined $\overline{\omega}_{mn} = \tilde{\omega}_{nm}$. Hence the expression of Eq. (4.32a) for Z_G becomes

$$Z_G = \int d\mu\, d\overline{\omega}\, d\omega\, \mathcal{G}[A_\ell]\exp\left[-\int d^3x \operatorname{Tr}\right.$$

$$\left. \times\left(\frac{1}{2\xi}(\sum_\ell L^\ell A_\ell)^2 - \overline{\omega}(\vec{x})\sum_\ell L^\ell D_\ell \omega(\vec{x})\right)\right]. \qquad (4.35a)$$

An alternative way of writing Eq. (4.35a), that is convenient for exhibiting the BRST invariance, is to introduce an auxiliary self-adjoint matrix field h and to reexpress Eq. (4.35a) as

$$Z_G = \int d\mu\, dh\, d\overline{\omega}\, d\omega\, \mathcal{G}[A_\ell]\exp\left[-\int d^3x \operatorname{Tr}\right.$$

$$\left. \times\left(\frac{\xi}{2}h^2 + ih\sum_\ell L^\ell A_\ell - \overline{\omega}(\vec{x})\sum_\ell L^\ell D_\ell \omega(\vec{x})\right)\right]. \qquad (4.35b)$$

Starting from Eq. (4.35b), we can now show that Z_G has a BRST invariance of the familiar form. Let θ be an $\vec{x}$-independent real c-number Grassmann parameter (i.e., a real 1×1 Grassmann matrix), and consider the variations defined by

$$\delta\omega = \omega^2\theta, \quad \delta\bar{\omega} = -ih\theta, \quad \delta h = 0,$$
$$\delta A_\ell = D_\ell\omega\theta, \quad \delta\chi = [\chi, \omega\theta] = \{\chi, \omega\}\theta. \tag{4.36}$$

Letting * denote complex conjugation, since

$$(\omega^2)^*_{ji} = \sum_\ell \omega^*_{j\ell}\omega^*_{\ell i} = \sum_\ell \omega_{\ell j}\omega_{i\ell} = -\sum_\ell \omega_{i\ell}\omega_{\ell j} = -(\omega^2)_{ij}, \tag{4.37a}$$

or in brief, $(\omega^2)^\dagger = -(\omega^\dagger)^2 = -\omega^2$, the property that ω is anti-self-adjoint is preserved by Eq. (4.36). Also, Eq. (4.36) leaves $\bar{\omega} = \tilde{\omega}^T$ anti-self-adjoint because h is self-adjoint.

We next show that Eq. (4.36) defines a nilpotent transformation, in the sense that the second variations of all quantities are zero. To verify this, we show that the variations of ω^2 and $D_\ell\omega$, which are the coefficients of θ on the right-hand side of the first and second lines of Eq. (4.36), are zero. (The variation of h on the right-hand side of the third line of Eq. (4.36) is already seen to be zero by the fourth line.) Evaluating the second variations of the first two lines, by again using Eq. (4.36), we have

$$\delta\omega^2 = \{\delta\omega, \omega\} = \{\omega^2\theta, \omega\} = \omega^2\{\omega, \theta\} = 0,$$
$$\delta D_\ell\omega = [\delta A_\ell, \omega] + D_\ell\delta\omega = [D_\ell\omega\theta, \omega] + D_\ell\omega^2\theta$$
$$= -\{D_\ell\omega, \omega\}\theta + \{D_\ell\omega, \omega\}\theta = 0,$$
$$\delta\{\chi, \omega\} = \{\delta\chi, \omega\} + \{\chi, \delta\omega\} = \{[\chi, \omega\theta], \omega\} + \{\chi, \omega^2\theta\} = 0. \tag{4.37b}$$

To see that Z_G is invariant, we note that the action on A_ℓ and χ of the BRST transformation of Eq. (4.36) is just a gauge transformation (albeit with a Grassmann valued generator $\Lambda = \omega\theta$, which is anti-self-adjoint since $(\omega\theta)^\dagger = -\theta^\dagger\omega^\dagger = \theta\omega = -\omega\theta$), and so the factors $d\mu$ and $\mathcal{G}[A_\ell]$ are invariant. The measure dh is trivially invariant, and the measure $d\bar{\omega}$ is invariant because $\delta\bar{\omega}$ has no dependence on $\bar{\omega}$. Using

$$\delta d\omega_{mn} = d(\delta\omega)_{mn} = d(\omega^2\theta)_{mn} = (\omega d\omega + d\omega\omega)_{mn}\theta, \tag{4.37c}$$

we have

$$\delta d\omega_{mn} = (\omega_{mm}d\omega_{mn} + d\omega_{mn}\omega_{nn})\theta + \ldots = d\omega_{mn}(\omega_{nn} - \omega_{mm})\theta + \ldots \tag{4.37d}$$

with $\ldots$ denoting terms that contain matrix elements $d\omega_{m'n'}$ with $(m', n') \neq (m, n)$. Consequently the Jacobian of transformation for $d\omega$ differs from unity by a term proportional to

$$\sum_{nm}(\omega_{nn} - \omega_{mm})\theta = 0, \tag{4.37e}$$

and so the measure $d\omega$ is also invariant. To complete the demonstration that Z_G is BRST invariant, we have to show that the gauge fixing part of the Hamiltonian

$$\mathbf{H}_G \equiv \int d^3x \, \text{Tr} \left(\frac{\xi}{2} h^2 + i h \sum_\ell L^\ell A_\ell - \bar{\omega} \sum_\ell L^\ell D_\ell \omega \right), \tag{4.38a}$$

is BRST invariant. Since we have already seen that $D_\ell \omega$ is invariant, and since h is trivially invariant, we have only to verify that

$$0 = \int d^3x \, \text{Tr} \left[i h \sum_\ell L^\ell \delta A_\ell - (\delta \bar{\omega}) \sum_\ell L^\ell D_\ell \omega \right] = \int d^3x \, \text{Tr} \, i h \sum_\ell L^\ell D_\ell \{\omega, \theta\}, \tag{4.38b}$$

which checks, completing the demonstration of BRST invariance of the generalized partition function.

4.5 Reduction of the Hilbert space modulo i_{eff}

Our aim in this section is to further study the structure of averages of dynamical variables over the canonical ensemble, and more specifically to study the implications of the fact that the canonical ensemble only partially breaks the originally assumed global unitary invariance group. This analysis is needed in Section 5.3 to establish the connection between vectors and operators in the underlying Hilbert space, and vectors and operators in what will become the emergent physical quantum mechanical Hilbert space.

We have seen in Section 4.2 that the canonical ensemble introduces an effective imaginary unit operator i_{eff} through

$$\langle \tilde{C} \rangle_{\text{AV}} = i_{\text{eff}} D_{\text{eff}}, \tag{4.39a}$$

where D_{eff} is assumed to be a real constant times the unit operator, and that the ensemble parameter $\tilde{\lambda}$ is functionally related to $\langle \tilde{C} \rangle_{\text{AV}}$ using only c-number coefficients. This means that the traceless, anti-self-adjoint parameter $\tilde{\lambda}$ must have the form

$$\tilde{\lambda} = \lambda i_{\text{eff}}, \tag{4.39b}$$

with λ a real c-number. Therefore if U_{eff} is a unitary matrix that commutes with i_{eff}

$$U_{\text{eff}}^\dagger U_{\text{eff}} = U_{\text{eff}} U_{\text{eff}}^\dagger = 1, \quad [U_{\text{eff}}, i_{\text{eff}}] = 0, \tag{4.39c}$$

then U_{eff} also commutes with $\tilde{\lambda}$

$$[U_{\text{eff}}, \tilde{\lambda}] = 0 \Rightarrow U_{\text{eff}} \tilde{\lambda} U_{\text{eff}}^\dagger = \tilde{\lambda} \tag{4.39d}$$

As a consequence, the canonical ensemble partially respects the assumed global unitary invariance of the dynamics: the integration measure $d\mu$, the trace Hamiltonian $\mathbf{H}$, and the trace quantity $\mathbf{N}$ are all invariant under general global unitary transformations of the matrix dynamical variables (cf. Eq. (2.7b) and the discussion of Section 4.1), but as we shall see in detail, the term in the exponent in the canonical ensemble $\text{Tr}\tilde{\lambda}\tilde{C}$ is invariant only under the subset U_{eff} of global unitary transformations that commute with i_{eff}. This has important consequences that we shall explore in this section and the next. We shall develop a formalism for isolating the effects of the residual global unitary invariance, and, after establishing that it is necessary to break this invariance in order to extract the full implications of the canonical ensemble, we shall give an explicit method for breaking the residual invariance by modifying the operator phase space measure.

Introducing the standard 2×2 Pauli matrices τ_1, τ_2, τ_3, it is convenient to rewrite Eq. (4.11c) in the form

$$i_{eff} = i\tau_3 1_K, \tag{4.40}$$

with 1_K a $K \times K$ unit matrix, where we recall that we have taken the dimension of the underlying matrix Hilbert space to be $N = 2K$. Letting $\tau_0 = 1_2$ denote the 2×2 unit matrix corresponding to the Pauli matrices $\tau_{1,2,3}$, a general $N \times N$ matrix M can be decomposed in the form

$$M = \frac{1}{2}(\tau_0 + \tau_3)M_+ + \frac{1}{2}(\tau_0 - \tau_3)M_- + \tau_1 M_1 + \tau_2 M_2, \tag{4.41a}$$

with $M_{+,-,1,2}$ four $K \times K$ matrices that operate on an independent Hilbert space from that acted on by the Pauli matrices $\tau_{1,2,3}$. (In other words, the Ms act as unit matrices on the 2-dimensional Hilbert space acted on by the Pauli matrices, while the Pauli matrices act as unit matrices on the K-dimensional Hilbert space acted on by the Ms.) Thus, corresponding to $M = i_{eff}$, we would have $M_+ = -M_- = i1_K$, $M_{1,2} = 0$. For general M, let us define M_{eff} and M_{12} by

$$M_{eff} = \frac{1}{2}(M - i_{eff}M i_{eff}) = \frac{1}{2}(\tau_0 + \tau_3)M_+ + \frac{1}{2}(\tau_0 - \tau_3)M_-,$$
$$M_{12} = M - M_{eff} = \tau_1 M_1 + \tau_2 M_2, \tag{4.41b}$$

so that M_{eff} and M_{12} give, respectively, the parts of M that commute and anticommute with i_{eff}

$$i_{eff}M_{eff} = M_{eff}i_{eff},$$
$$i_{eff}M_{12} = -M_{12}i_{eff}. \tag{4.41c}$$

Combining Eqs. (4.41b,c), we get the useful relation

$$2i_{eff}M_{eff} = \{i_{eff}, M\}. \tag{4.41d}$$

We see that for the subset of matrix operators M_{eff} that commute with i_{eff}, the original N-dimensional Hilbert space diagonalizes into two subspaces of dimension K, on the first of which i_{eff} acts as $i1_K$ and M_{eff} acts as M_+, and on the second of which i_{eff} acts as $-i1_K$ and M_{eff} acts as M_-. In Chapter 5 we shall identify these two K-dimensional subspaces with the physical subspace and with a second, complex conjugated copy of the physical subspace, respectively.

Using this notation, let us examine the unitary transformation behavior of the term $\text{Tr}\tilde{\lambda}\tilde{C}$ in the partition function. Substituting Eq. (4.39b), we have

$$\text{Tr}\tilde{\lambda}\tilde{C} = \lambda \text{Tr}\, i_{\text{eff}}\tilde{C}. \tag{4.42a}$$

Under a general unitary transformation of the dynamical variables, we have

$$q_r \to U^\dagger q_r U, \quad p_r \to U^\dagger p_r U, \quad \tilde{C} \to U^\dagger \tilde{C} U, \tag{4.42b}$$

and so the right-hand side of Eq. (4.42a) becomes

$$\lambda \text{Tr}\, U i_{\text{eff}} U^\dagger \tilde{C}, \tag{4.42c}$$

which in general differs from Eq. (4.42a) because a general U does not commute with i_{eff}. Thus the $\text{Tr}\tilde{\lambda}\tilde{C}$ term in the canonical ensemble breaks the global unitary invariance of the underlying dynamics. However, when U in Eq. (4.42b) is restricted to have the structure U_{eff} that commutes with i_{eff}, the transformation of Eq. (4.42b) is still an invariance of Eq. (4.42a), and hence is an invariance of the canonical ensemble.

In group theoretic terms (Adler and Kempf, 1998), the original $U(N)$ global unitary invariance of $\mathbf{H}$ is broken, by the term $\text{Tr}\tilde{\lambda}\tilde{C}$ in the canonical ensemble, to $U(K) \times U(K) \times R$, with R the discrete reflection symmetry that interchanges the eigenvalues $\pm i$ of i_{eff}. This is clearly the largest symmetry group of the ensemble for which one can have $\langle\tilde{C}\rangle_{AV} \neq 0$. If one were to attempt to preserve the full $U(N)$ symmetry by taking an ensemble with $\tilde{\lambda} = i\lambda$, with λ a c-number, then in the canonical ensemble the term $\text{Tr}\tilde{\lambda}\tilde{C}$ would vanish by virtue of the tracelessness of $\tilde{C}$, and the resulting ensemble would have $\langle\tilde{C}\rangle_{AV} = 0$. Requiring the largest possible nontrivial symmetry group plays the role in our derivation of making the emergent Planck constant a c-number; if, however, we were to sacrifice all of the $U(N)$ symmetry by allowing a generic $\tilde{\lambda}$, then the emergent canonical commutation relations derived in Section 5.3 below would generically yield a matrix $\hbar$ acting nontrivially on the states of Hilbert space, which would be inconsistent with an emergent Heisenberg dynamics. It would clearly be desirable to have a deeper justification from first principles of our choice of canonical ensemble, perhaps based on a more detailed understanding of the underlying trace dynamics, but at present we must simply introduce it as a postulate.

The residual unitary invariance of the canonical ensemble has the following consequence. Let us write the integration measure $d\mu$ as

$$d\mu = d[U_{\text{eff}}]d\hat{\mu}, \tag{4.43a}$$

with $d[U_{\text{eff}}]$ the Haar measure for integration over the subgroup of global unitary transformations U_{eff} that commute with i_{eff}, and with $d\hat{\mu}$ the integration measure over the operator phase space subject to the restriction that an overall global unitary transformation U_{eff} is kept fixed. Let us consider the canonical ensemble average of a polynomial operator R_{eff}, that is a function (with c-number coefficients) of i_{eff} and of the underlying dynamical variables q_r, p_r and that commutes with i_{eff}

$$R_{\text{eff AV}} \equiv \langle R_{\text{eff}} \rangle_{\text{AV}} = \frac{\int d[U_{\text{eff}}]d\hat{\mu}\rho R_{\text{eff}}}{\int d[U_{\text{eff}}]d\hat{\mu}\rho}. \tag{4.43b}$$

We can relate the general operator variables q_r, p_r to operator variables $\hat{q}_r$, $\hat{p}_r$ that have an overall U_{eff} rotation frozen, by writing

$$q_r = U_{\text{eff}}^{\dagger}\hat{q}_r U_{\text{eff}}, \quad p_r = U_{\text{eff}}^{\dagger}\hat{p}_r U_{\text{eff}}, \tag{4.43c}$$

and so correspondingly we have

$$R_{\text{eff}} = U_{\text{eff}}^{\dagger}\hat{R}_{\text{eff}}U_{\text{eff}}, \tag{4.43d}$$

with $\hat{R}_{\text{eff}}$ obtained from R_{eff} by the replacements $q_r, p_r \to \hat{q}_r, \hat{p}_r$. Since the canonical ensemble is invariant under unitary transformations that commute with i_{eff}, we have $\rho = \hat{\rho}$, with $\hat{\rho}$ constructed in the same manner as ρ, but using the variables $\hat{q}_r$, $\hat{p}_r$ in place of q_r, p_r. Putting all these ingredients together, we can rewrite Eq. (4.43b) in the form

$$R_{\text{eff AV}} = \frac{\int d[U_{\text{eff}}]U_{\text{eff}}^{\dagger}R_{\text{eff A}\hat{\text{V}}}U_{\text{eff}}}{\int d[U_{\text{eff}}]}, \tag{4.44a}$$

with $R_{\text{eff A}\hat{\text{V}}}$ given by

$$R_{\text{eff A}\hat{\text{V}}} \equiv \frac{\int d\hat{\mu}\hat{\rho}\hat{R}_{\text{eff}}}{\int d\hat{\mu}\hat{\rho}}. \tag{4.44b}$$

Writing all matrix quantities in terms of $+$ and $-$ components according to Eq. (4.41b), Eqs. (4.44a,b) separate into the independent $\pm$ components

$$R_{\text{eff AV}\pm} = \frac{\int d[U_{\text{eff}\pm}]U_{\text{eff}\pm}^{\dagger}R_{\text{eff A}\hat{\text{V}}\pm}U_{\text{eff}\pm}}{\int d[U_{\text{eff}\pm}]}, \tag{4.44c}$$

with $R_{\text{eff}\,\hat{A}\hat{V}\pm}$ given by

$$R_{\text{eff}\,\hat{A}\hat{V}\pm} \equiv \frac{\int d\hat{\mu}\,\hat{\rho}\,\hat{R}_{\text{eff}\pm}}{\int d\hat{\mu}\,\hat{\rho}}. \qquad (4.44\text{d})$$

In both the $\pm$ cases, the integral on the right-hand side of Eq. (4.44c) has the general form

$$I[M_K] = \frac{\int d[U_K]U_K^\dagger M_K U_K}{\int d[U_K]}, \qquad (4.45\text{a})$$

with M_K a $K \times K$ matrix and with U_K a $K \times K$ unitary matrix. But replacing M_K by $V_K^\dagger M_K V_K$, with V_K unitary, and using the invariance property $d[V_K U_K] = d[U_K]$ of the Haar measure, we see that $I[M_K] = I[V_K^\dagger M_K V_K]$ for arbitrary unitary V_K. Thus $I[M_K]$ is a linear, unitary invariant function of M_K, which on the unit matrix takes the value $I[1_K] = 1_K$. These properties imply that $I[M_K]$ is given by the trace

$$I[M_K] = \frac{1}{K}1_K \operatorname{Tr}_K M_K. \qquad (4.45\text{b})$$

We learn from this that if we take an unrestricted average of R_{eff} over the canonical ensemble, the interesting matrix operator structure is averaged out. To preserve this structure, we must restrict the integration in the canonical ensemble to leave an overall global unitary transformation U_{eff} fixed, as in $R_{\text{eff}\,\hat{A}\hat{V}}$ of Eq. (4.44b). A second, and perhaps deeper reason for requiring a global unitary fixing in the $i_{\text{eff}} = \pm i$ subspaces is that, as we shall see in Section 5.3, Poincaré transformations and internal symmetry transformations in the emergent quantum theory both take the form of global unitary transformations U_{eff}. Therefore we must completely break this residual unitary invariance group in the canonical ensemble, in order for the ensemble not to inadvertently include integrations over the symmetry groups of the emergent theory. Since introducing a global unitary fixing introduces arbitrary parameters into the unitary fixed ensemble (just as introducing a gauge fixing introduces arbitrary parameters in gauge theories), we must make sure that physical observables in the emergent quantum theory correspond to quantities that are independent of the choice of unitary fixing. We shall see in Section 4.6 that trace quantities are always independent of the unitary fixing (i.e., they are the analogs of gauge invariants in gauge theories), and we shall show in Section 5.3 that observable matrix elements in the emergent quantum theory correspond to trace quantities in the underlying trace dynamics.

In practice, it is actually sufficient to fix only the $SU(K)$ subgroup of the $U(K)$ group, since, when U_K in Eq. (4.45a) is an overall phase times the unit matrix 1_K, it commutes with M_K and does not wash out the matrix structure. The explicit construction given in Section 4.6 involves such an $SU(K)$ fixing. Henceforth, we will use the term unitary fixing to mean either a fixing of the groups $U(K)$ in the $i_{eff} = \pm i$ sectors, or a fixing of their $SU(K)$ subgroups. In constructing a unitary fixing, we shall always make the simplifying assumption that the structure of the trace Hamiltonian is such that we cannot split the dynamical variables $\{x_r\}$ into two disjoint sets $\{x_r^I\}$ and $\{x_r^{II}\}$, for which the trace Hamiltonian *exactly* separates into disjoint pieces

$$\mathbf{H} = \mathbf{H}^I[\{x_r^I\}] + \mathbf{H}^{II}[\{x_r^{II}\}]. \tag{4.46}$$

If Eq. (4.46) were to hold, the fact that $\tilde{C}$ and $\mathbf{N}$ are additive over the dynamical degrees of freedom would then imply exact factorization of the partition function Z according to $Z = Z^I Z^{II}$, and we would then have to address the same problem of fixing a global unitary invariance at the level of both Z^I and Z^{II}. Put another way, we shall assume in the analysis of the next section that our trace dynamics is irreducible, in the sense that it cannot be exactly reduced to two or more independent trace dynamics systems. Once this assumption has been made, it suffices to fix a global unitary rotation of any *one* canonical pair of dynamical variables x_R, x_{R+1}, which we shall assume to be self-adjoint bosonic variables (and which we shall denote by A, B in the more detailed discussion of the next section).

In our applications of the restricted measure we will need to know the restricted canonical average of $\tilde{C}$. Since the unitary fixing acts within the subspaces on which i_{eff} is diagonal, and does not mix the $i_{eff} = \pm i$ subspaces, we have the unitary fixed analog of Eq. (4.10c)

$$[\tilde{\lambda}, \langle \mathcal{O} \rangle_{\widehat{AV}}] = 0, \tag{4.47a}$$

for any operator $\mathcal{O}$ constructed from the $\{x_r\}$ using only c-number coefficients. Thus, for $\mathcal{O} = \tilde{C}$, the expectation $\langle \tilde{C} \rangle_{\widehat{AV}}$ has the general form

$$\langle \tilde{C} \rangle_{\widehat{AV}} = i_{eff}\hbar + \frac{1}{2}(\tau_0 + \tau_3)\Delta_+ + \frac{1}{2}(\tau_0 - \tau_3)\Delta_-, \tag{4.47b}$$

with Δ_+ and Δ_- anti-self-adjoint traceless $K \times K$ matrices. The detailed structure of $\Delta_\pm$ depends on the unitary fixing, and there is no general reason for them to vanish. Motivated by the extensive form of $\tilde{C}$, and the fact that the unitary fixing is implemented on a single canonical pair, we shall assume, in the limit when the number of canonical pairs is very large, that the matrices $\Delta_\pm$ are small, and can be neglected relative to the term $i_{eff}\hbar$ that is independent of the choice of unitary fixing. This will be included among our assumptions/approximations in Section 5.3, where we give conditions for emergence of an effective quantum mechanical structure.

4.6 Global unitary fixing*

Let us now give an explicit method for global unitary fixing, following the analysis of Adler and Horwitz (2003). We work now in one of the two K-dimensional subspaces of the original $N = 2K$-dimensional Hilbert space, on which i_{eff} acts as a multiple of the unit matrix, so that there is a residual unbroken $U(K)$ invariance group. Typically, in matrix model calculations, an overall global unitary invariance in the partition function is partially integrated out as a first step, thus eliminating a $U(K)/U(1)^K$ subgroup of the global unitary group. In this section we shall proceed in an alternative fashion, by using the De Witt–Faddeev–Popov framework to impose a set of unitary invariance fixing conditions, that completely break the $SU(K)$ subgroup of the global unitary invariance group $U(K)$. One can think of this construction as a type of polar decomposition, based on computing the residue modulo the action of the $SU(K)$ subgroup. This allows one to define matrix-valued correlation functions, which give additional structural information about the system, but which (like gauge potentials in gauge field theory) depend on the choice of unitary fixing.

We shall change notation for the underlying matrix variables in this section, to follow that customary in the literature on matrix models (see, e.g., Brézin and Wadia, 1993), and to keep the discussion of this section self-contained, shall repeat a few of the results of the preceding section. We shall focus exclusively on the bosonic sector of the theory, since the inclusion of fermions involves no further complications. Let $M_1, \ldots, M_D$ be a set of $K \times K$ complex self-adjoint matrices (note that in this section D plays the role played by $2n_B$ earlier), and let $\mathbf{S} = \text{Tr}S(M_1, \ldots, M_D)$ be an "energy" functional which is constructed using only c-number coefficients, from which we form a partition function Z defined by

$$Z = \int dM \exp(-\mathbf{S}). \tag{4.48a}$$

Here we have written

$$dM = \prod_{d=1}^{D} d[M_d], \tag{4.48b}$$

and the integration measure $d[M]$ for the self-adjoint matrix M is defined in terms of the real and imaginary parts of the matrix elements M_{ij} of M by the measure introduced in Section 4.1,

$$d[M] = \prod_{i} dM_{ii} \prod_{i<j} d\text{Re}M_{ij} d\text{Im}M_{ij}. \tag{4.48c}$$

As we have already noted, the measure $d[M]$ is unitary invariant, in other words, if U is a fixed $K \times K$ unitary matrix, then

$$d[U^\dagger M U] = d[M]. \tag{4.49a}$$

If we make the same unitary transformation U on all of the matrices $M_d, \quad d = 1, \ldots, D$, then by our assumption that S involves no fixed matrix coefficients, $\mathbf{S}$ is invariant by virtue of the cyclic property of the trace

$$\mathbf{S}[\{U^\dagger M U\}] = \mathbf{S}[\{M\}], \tag{4.49b}$$

and so the partition function Z has a global unitary invariance.

As we have seen in the preceding section, the global unitary invariance of Z must be taken into account in calculating correlations of the various matrices M_d averaged over the partition function. Let $Q[\{M\}]$ be an arbitrary polynomial in the matrices $M_1, \ldots, M_D$ constructed using only c-number coefficients, so that under global unitary transformations, Q transforms as

$$Q[\{U^\dagger M U\}] = U^\dagger Q[\{M\}]U. \tag{4.50a}$$

Correspondingly, let

$$\mathbf{Q} = \mathrm{Tr} Q, \tag{4.50b}$$

so that $\mathbf{Q}$ is a global unitary invariant. One can now consider the calculation of averages of $\mathbf{Q}$ and of Q respectively over the ensemble defined by Eq. (4.48a). In the case of the trace polynomial $\mathbf{Q}$ one has

$$\langle \mathbf{Q} \rangle_{\mathrm{AV}} = Z^{-1} \int dM \exp(-\mathbf{S}) \mathbf{Q}, \tag{4.51a}$$

which, because of the global unitary invariance, involves an overall structure-independent unitary integration that is typically done as the first step by using the change of variables given by Mehta (1967) for one of the matrix arguments on which $\mathbf{Q}$ depends. Let us now consider the corresponding average of the polynomial Q over the ensemble

$$\langle Q \rangle_{\mathrm{AV}} = Z^{-1} \int dM \exp(-\mathbf{S}) Q. \tag{4.51b}$$

Making a global unitary transformation on all of the matrix integration variables, and using the invariance of dM and of $\mathbf{S}$ and the covariance of Q given in Eq. (4.50a), we then find that

$$\langle Q \rangle_{\mathrm{AV}} = U^\dagger \langle Q \rangle_{\mathrm{AV}} U, \tag{4.52a}$$

for all unitary matrices U. Thus by Schur's lemma (which applies since $U(K)$ acts irreducibly on the complex K-dimensional vector space) $\langle Q \rangle_{\text{AV}}$ must be a c-number multiple of the unit matrix, so that by taking the trace, we learn that

$$\langle Q \rangle_{\text{AV}} = K^{-1} \langle \mathbf{Q} \rangle_{\text{AV}}, \qquad (4.52b)$$

and all nontrivial matrix information (e.g., the unitary orientation and nontrivial operator properties) contained in Q has been lost. (Comparing with the discussion of Section 4.5, Eq. (4.52b) is a rewriting of Eq. (4.45b) in the notation of the present section.)

In order to retain access to the matrix information contained in Q, let us now proceed in an alternative fashion. Let us define a measure $\hat{d}M$ in which the $SU(K)$ subgroup of the global unitary invariance group has been fixed. (The full global unitary invariance group is the product of this $SU(K)$ with a global $U(1)$ that is an overall phase times the unit matrix; since this $U(1)$ commutes with Q, averaging over it causes no loss of the matrix information contained in Q, and so fixing the overall $U(1)$ is not necessary.) We then define the average of Q over the unitary fixed ensemble as

$$\langle Q \rangle_{\hat{\text{AV}}} = Z^{-1} \int \hat{d}M \exp(-\mathbf{S}) Q, \qquad (4.53a)$$

now with

$$Z = \int \hat{d}M \exp(-\mathbf{S}) \qquad (4.53b)$$

the partition function in which the global unitary invariance has been broken, and an orientation on the K-dimensional vector space has been fixed. Clearly, the procedure just described is a global unitary analog of the gauge fixing customarily employed in the case of local gauge invariances. If we change the recipe for fixing the global unitary invariance, then the average defined by Eq. (4.53a) will change in a manner that is in general complicated. However, we will show that the average of $\mathbf{Q}$ in the unitary fixed ensemble is independent of the fixing and is equal to that defined in Eq. (4.51a) by averaging over the original ensemble, so that

$$\langle \mathbf{Q} \rangle_{\hat{\text{AV}}} = \langle \mathbf{Q} \rangle_{\text{AV}}. \qquad (4.54)$$

In other words, the average of the trace of Q takes the same value for any choice of unitary fixing. To make an analogy with local gauge fixing in gauge theories, the trace polynomials $\mathbf{Q}$ are analogs of gauge invariant functions, while polynomials Q without a trace are analogs of gauge variant quantities. Just as the gauge variant potentials contain useful information in gauge theories, the unitary fixing-variant averages of polynomials Q contain useful structural information about matrix models.

To prove Eq. (4.54), we proceed by analogy with the standard De Witt–Faddeev–Popov procedure used for local gauge fixing. Let us write an infinitesimal $SU(K)$ transformation in generator form as $U = \exp(G)$, with G anti-self-adjoint and traceless. We take as the $K^2 - 1$ infinitesimal parameters of the $SU(K)$ transformation the real numbers g_j, $j = 1, \ldots, K^2 - 1$, with those for $j = 1, \ldots, K(K - 1)$ given by the real and imaginary parts of the off-diagonal matrix elements of G, that is, by $\mathrm{Re}\, G_{ij}$ and $\mathrm{Im}\, G_{ij}$ for $i < j$. The remaining ones for $j = K(K - 1) + 1, \ldots, K^2 - 1$ are given by the differences of the imaginary parts of the diagonal matrix elements of G, that is, by $\mathrm{Im}(G_{11} - G_{22}), \ldots, \mathrm{Im}(G_{11} - G_{KK})$. Let $f_j(\{M\})$, $j = 1, \ldots, K^2 - 1$ be a set of functions of the matrices $M_1, \ldots, M_D$ with the property that the equations $f_j(\{M\}) = 0$, $j = 1, \ldots, K^2 - 1$ completely break the $SU(K)$ invariance group, so that the only solution of $f_j(\{M + [G, M]\}) = 0$, $j = 1, \ldots, K^2 - 1$ is $g_j = 0$, $j = 1, \ldots, K^2 - 1$. We consider now the integral

$$\mathcal{J} = \int dM \mathcal{G}[\{M\}] \mathcal{K}[\{f_j(M)\}] \det \left(\frac{\partial f_i(\{M + [G, M]\})}{\partial g_j} \bigg|_{G=0} \right), \qquad (4.55a)$$

with the function $\mathcal{K}[\{f_j(M)\}]$ taken as

$$\mathcal{K}[\{f_j(M)\}] = \prod_{j=1}^{K^2-1} \delta(f_j). \qquad (4.55b)$$

Here $\mathcal{G}$ is a global unitary invariant function of the matrices $M_1, \ldots, M_D$, such as a trace polynomial $\mathbf{Q}$ or any function of trace polynomials (for example the statistical ensemble weight $\exp(-\mathbf{S})$). Equation (4.55a) has the standard form of the De Witt–Faddeev–Popov analysis, as formulated for example in the text of Weinberg (1996) (except that when dealing with a non-compact local gauge invariance, where the limits of integration lie at infinity, one can take the function $\mathcal{K}$ to be a general function of gauge variant functions f_j; in the compact case considered here, the delta functions of Eq. (4.55b) must be used in order to make the integration limits irrelevant.) The standard De Witt–Faddeev–Popov argument then shows that the integral in Eq. (4.55a) is independent of the constraints f_j. Briefly, the argument proceeds by replacing the dummy variable of integration dM by dM^{V_K}, where $M^{V_K} = V_K^\dagger M V_K$, and integrating over the $SU(K)$ matrix V_K. The group property of unitary transformations together with the chain rule then converts the determinant in Eq. (4.55a) into a Jacobian transforming the V_K integration into an integration over the constraints f_j, permitting the delta functions in Eq. (4.55b) to be integrated to give unity. This shows that the result is independent of the constraints, and that it is the same as the result obtained by integrating over the original unfixed ensemble, thus establishing Eq. (4.54). Clearly, this argument works only when the function $\mathcal{G}$ is a unitary invariant, so that it has no dependence on V_K. For example, if $\mathcal{G}$ is replaced by a polynomial in the matrices *without* an overall trace, then the unitary fixing constraints cannot be eliminated by integrating over V_K, and the result depends on the unitary fixing in a complicated way.

A specific realization of the general unitary fixing can be given when $D \geq 2$, so that the set of matrices $M_1, \ldots, M_D$ contains at least two independent self-adjoint matrices $A = M_1$ and $B = M_2$. We take the functions f_j, $j = 1, \ldots, K^2 - 1$ to be linear functions of A and B, constructed as follows. As the f_j for $j = 1, \ldots, K(K-1)$ we take the real and imaginary parts of the off-diagonal matrix elements of A, that is, the functions $\text{Re} \, A_{ij}$ and $\text{Im} \, A_{ij}$ for $i < j$. Equating these functions to zero forces the matrix A to be diagonal. The K remaining diagonal unitary transformations then commute with A, so that no further conditions can be furnished by use of A alone. However, the diagonal $SU(K)$ transformations can always be used to make the off-diagonal matrix elements in the first row of the second matrix B have vanishing imaginary parts, leaving a residual Z_2^{K-1} symmetry that is broken by requiring these matrix elements to have positive semidefinite real parts. So for the remaining conditions f_j for $j = K(K-1) + 1, \ldots, K^2 - 1$, we take the $K - 1$ functions $\text{Im} \, B_{1j}$, $j > 1$, and we restrict the integrations over $\text{Re} \, B_{1j}$, $j > 1$ to run from 0 to ∞. Since the function $\mathcal{K}$ chosen in Eq. (4.55b) enforces the conditions $f_j = 0$ in a sharp manner, they can be used to simplify the expression for the De Witt–Faddeev–Popov determinant. A simple calculation now shows that when the f_j all vanish, the matrix elements of the commutator $[G, M]$ needed in Eq. (4.55a) are given by

$$\text{Re}[G, A]_{ij} = \text{Re} \, G_{ij}(A_{jj} - A_{ii}),$$
$$\text{Im}[G, A]_{ij} = \text{Im} \, G_{ij}(A_{jj} - A_{ii}), \tag{4.56a}$$
$$\text{Im}[G, B]_{1j} = \text{Re} \, B_{1j} \text{Im}(G_{11} - G_{jj}) + R,$$

with R a remainder containing only off-diagonal elements $G_{i \neq j}$ of the matrix G. Since Eq. (4.56a) shows that the matrix

$$\left(\frac{\partial f_i(\{M + [G, M]\})}{\partial g_j} \bigg|_{G=0} \right) \tag{4.56b}$$

is triangular (its upper off-diagonal matrix elements are all zero because R has no dependence on the diagonal matrix elements of G), its determinant is given by the product of its diagonal matrix elements. Thus we have

$$\Delta \equiv \det \left(\frac{\partial f_i(\{M + [G, M]\})}{\partial g_j} \bigg|_{G=0} \right) = \prod_{i<j}(A_{ii} - A_{jj})^2 \prod_{j=2}^{K} \operatorname{Re} B_{1j}, \quad (4.57a)$$

the first factor of which is the familiar squared Vandermonde determinant. Substituting Eqs. (4.55b) and (4.57a) into Eq. (4.55a), we thus arrive at the formula for the unitary fixed integral

$$\mathcal{J} = \int \prod_{d=3}^{D} d[M_d] \left(\prod_{i=1}^{K} dA_{ii} dB_{ii} \right) \left(\prod_{j=2}^{K} d\operatorname{Re} B_{1j} \right)$$

$$\times \left(\prod_{2 \le i < j} d\operatorname{Re} B_{ij} d\operatorname{Im} B_{ij} \right) \Delta \mathcal{G}[\{M\}], \quad (4.57b)$$

with the integrals over $\operatorname{Re} B_{1j}$, $j = 2, \ldots, K$ running over positive values only. The part of this analysis involving only a single matrix A is well-known in the literature (see Wadia, 1981; Yoneya and Itoyama, 1982; and Coleman, 1989); what has been added here is the complete $SU(K)$ fixing obtained by imposing a condition on a second matrix B as well. The part of Eqs. (4.57a,b) involving each B_{1j} is just a planar radial integral $\int_0^\infty \rho d\rho$, with $\rho = |B_{1j}| = \operatorname{Re} B_{1j}$, where the associated angular integral $\int_0^{2\pi} d\phi$ has been omitted because it corresponds to a $U(1)$ factor that has been fixed by the condition $\phi = 0$.

With this choice of unitary fixing, the unitary fixed average $\hat{Q} \equiv \langle Q \rangle_{\hat{A}V}$ defined in Eq. (4.53a) has a characteristic form that is dictated by the symmetries of the unitary fixed ensemble. Since the unitary fixing conditions are symmetric under permutation of the basis states with labels $2, 3, \ldots, K$, and since this permutation is also a symmetry of the unfixed measure dM, the matrix $\hat{Q}$ must be symmetric under this permutation of basis states. Thus, there are only five independent matrix elements

$$\hat{Q}_{11} = \alpha,$$

$$\hat{Q}_{jj} = \beta, \ j = 2, \ldots, K,$$

$$\hat{Q}_{1j} = \gamma, \ j = 2, \ldots, K, \quad (4.57c)$$

$$\hat{Q}_{i1} = \delta, \ i = 2, \ldots, K,$$

$$\hat{Q}_{ij} = \epsilon, \ 2 \le i \ne j \le K.$$

In this notation, the original unfixed average $\bar{Q} \equiv \langle Q \rangle_{AV}$ defined by Eq. (4.52b) is given by

$$\bar{Q} = K^{-1}\mathrm{Tr}\hat{Q} = K^{-1}[\alpha + (K-1)\beta], \tag{4.57d}$$

showing explicitly that there is a loss of structural information in using the unfixed average. But even the unitary fixed average has a structure that is greatly restricted as compared with a general $K \times K$ matrix. (Similar reasoning applies to the partial unitary fixing in which one only imposes the condition that A should be diagonal. Since this condition is symmetric under permutation of the basis states with labels $1, \ldots, K$, the partially unitary fixed average of a polynomial Q defined by integrating with the measure $(\prod_{i=1}^{K} dA_{ii})(\prod_{i<j}(A_{ii} - A_{jj})^2)\prod_{d=2}^{D} d[M_d]$ must also have this permutation symmetry, and thus must be a c-number times the unit matrix.)

We now proceed to rewrite the unitary fixed measure in a form that allows us to determine the general form of Ward identities, and (in Appendix G) to identify an associated BRST invariance transformation. We begin by introducing ghost integrals to represent the determinant Δ. Let ω_{ij} and $\tilde{\omega}_{ij}$ be the matrix elements of independent $K \times K$ complex anti-self-adjoint Grassmann matrices ω and $\tilde{\omega}$. We take ω to be traceless, $\mathrm{Tr}\omega = 0$, while we take $\tilde{\omega}$ to have a vanishing 11 matrix element, $\tilde{\omega}_{11} = 0$. The integration measure for ω is defined by

$$d\omega = \prod_{i<j} d\mathrm{Re}\omega_{ij} d\mathrm{Im}\omega_{ij} \prod_{j=2}^{K} d\mathrm{Im}(\omega_{jj} - \omega_{11}), \tag{4.58a}$$

while the integration measure for $\tilde{\omega}$ is taken as

$$d\tilde{\omega} = \prod_{i<j} d\mathrm{Re}\tilde{\omega}_{ij} d\mathrm{Im}\tilde{\omega}_{ij} \prod_{j=2}^{K} d\mathrm{Im}\tilde{\omega}_{jj}. \tag{4.58b}$$

We can now use these Grassmann matrices to give a ghost representation of the factors in Eq. (4.57a) involving the matrices A and B. Since the matrix A is diagonal, we have

$$\mathrm{Tr}\tilde{\omega}[\omega, A] = \sum_{i \neq j} \tilde{\omega}_{ji}(A_{ii} - A_{jj})\omega_{ij}. \tag{4.59a}$$

Hence up to an overall sign, the square of the Vandermonde determinant $\prod_{i<j}(A_{ii} - A_{jj})^2$ is given by the ghost integral

$$\int d'\omega \, d'\tilde{\omega} \exp(\mathrm{Tr}\tilde{\omega}[\omega, A]), \tag{4.59b}$$

with the diagonal factors $d\mathrm{Im}(\omega_{jj} - \omega_{11})$, $d\mathrm{Im}\tilde{\omega}_{jj}$, $j = 2, \ldots, K$ omitted from the primed integration measures $d'\omega$ and $d'\tilde{\omega}$. To represent the second factor in

Eq. (4.57a) as a ghost integral, we use the diagonal matrix elements of ω and $\tilde{\omega}$ in an analogous fashion. Thus, up to a phase, the factor $\prod_{j=2}^{K} \operatorname{Re} B_{1j}$ is given by the ghost integral

$$\int \prod_{j=2}^{K} d\operatorname{Im}(\omega_{jj} - \omega_{11}) d\operatorname{Im}\tilde{\omega}_{jj} \exp\left(\sum_{j=2}^{K} \tilde{\omega}_{jj} (\operatorname{Re} B_{1j}) i (\omega_{jj} - \omega_{11}) \right). \quad (4.59c)$$

By defining a matrix X by $X_{11} = 0$; $X_{ij} = 0$, $2 \le i, j \le K$; $X_{1j} = X_{j1} = \frac{i}{2}\tilde{\omega}_{jj}(\omega_{jj} - \omega_{11})$, $j = 2, \ldots, K$, the exponent in Eq. (4.59c) can be written as $\operatorname{Tr} XB$, so that Eq. (4.59c) becomes

$$\int \prod_{j=2}^{K} d\operatorname{Im}(\omega_{jj} - \omega_{11}) d\operatorname{Im}\tilde{\omega}_{jj} \exp(\operatorname{Tr} XB). \quad (4.59d)$$

Combining Eqs. (4.58a,b) and (4.59b,c,d), we see that up to an overall phase the determinant Δ introduced in Eq. (4.57a) has the ghost representation

$$\Delta \propto \int d\omega d\tilde{\omega} \exp(\operatorname{Tr}\tilde{\omega}[\omega, A] + \operatorname{Tr} XB). \quad (4.60)$$

Our next step is to rewrite the product of δ functions in Eq. (4.55b) and the half-line restriction on the integrals over $\operatorname{Re} B_{1j}$ in terms of their Fourier representations, by introducing three sets of auxiliary variables. One set are the elements h_{ij} of a self-adjoint $K \times K$ matrix h with vanishing diagonal matrix elements, so that $h_{ii} = 0$, $i = 1, \ldots, K$. The integration measure for this set is defined as

$$dh = \prod_{i<j} d\operatorname{Re} h_{ij} d\operatorname{Im} h_{ij}. \quad (4.61a)$$

The second set are $K - 1$ real numbers H_j, $j = 2, \ldots, K$, with integration measure

$$dH = \prod_{j=2}^{K} dH_j. \quad (4.61b)$$

In terms of these variables, the product of δ functions of Eq. (4.55b) can be represented (up to an overall constant factor) as

$$\prod_{j=1}^{K^2-1} \delta(f_j) \propto \int dh dH \exp\left(i\operatorname{Tr} hA + i \sum_{j=2}^{K} H_j \operatorname{Im} B_{1j} \right). \quad (4.62a)$$

The third set are $K - 1$ complex numbers k_j, $j = 2, \ldots, K$, integrated along a contour on the real axis with integration measure

$$dk = \prod_{j=2}^{K} dk_j / (k_j - i\epsilon), \tag{4.62b}$$

with infinitesimal positive ϵ. These can be used to insert a product of step functions $\prod_{j=2}^{K} \theta(\mathrm{Re}\, B_{1j})$ into Eq. (4.57b)

$$\prod_{j=2}^{K} \theta(\mathrm{Re}\, B_{1j}) \propto \int dk \exp\left(i \sum_{j=2}^{K} k_j \mathrm{Re}\, B_{1j}\right), \tag{4.62c}$$

allowing the integrals over the $\mathrm{Re}\, B_{1j}$ in Eq. (4.57b) to be taken from $-\infty$ to ∞.

Defining a matrix Y by $Y_{11} = 0$; $Y_{ij} = 0$, $2 \le i, j \le K$; $Y_{1j} = -Y_{j1} = -\frac{1}{2}H_j$, the second term in the exponent in Eq. (4.62a) can be rewritten as $i \sum_{j=2}^{K} H_j \mathrm{Im}\, B_{1j} = \mathrm{Tr}\, Y B$, and so an alternative form of Eq. (4.62a) is

$$\prod_{j=1}^{K^2-1} \delta(f_j) \propto \int dh\, dH \exp(i\,\mathrm{Tr}\, hA + \mathrm{Tr}\, Y B). \tag{4.62d}$$

Similarly, defining a matrix V by $V_{11} = 0$; $V_{ij} = 0$, $2 \le i, j \le K$; $V_{1j} = V_{j1} = \frac{1}{2}ik_j$, the exponent in Eq. (4.62c) can be rewritten as $\sum_{j=2}^{K} ik_j \mathrm{Re}\, B_{1j} = \mathrm{Tr}\, V B$, and so an alternative form of Eq. (4.62c) is

$$\prod_{j=2}^{K} \theta(\mathrm{Re}\, B_{1j}) \propto \int dk \exp(\mathrm{Tr}\, V B). \tag{4.62e}$$

These equations allow us to write Eq. (4.57b) in terms of the unrestricted measure dM, and the ghost representation of Δ, as

$$\mathcal{J} = C \int dM\, dh\, dH\, dk\, d\omega\, d\tilde{\omega} \exp\left(i\,\mathrm{Tr}\, hA + \mathrm{Tr}\,\tilde{\omega}[\omega, A]\right.$$
$$\left. + \mathrm{Tr}(V + X + Y)B\right)\mathcal{G}[\{M\}], \tag{4.63}$$

with C an overall constant factor. This representation of $\mathcal{J}$ will be used to discuss Ward identities obeyed by the matrix-valued correlations, while an alternative form, given in Appendix G, will be used to establish a BRST invariance.

We proceed now to derive Ward identities from unitary fixed expectations of trace polynomials $\mathbf{Q}$; these Ward identities play a central role in the arguments for an emergent quantum theory given in Chapter 5. Employing the specific unitary fixing of Eq. (4.63) in the definition of Eqs. (4.53a,b), as applied to $\mathbf{Q} = \mathrm{Tr}\, Q$, and using the cyclic property of the trace to rewrite $\text{Tr}\tilde{\omega}[\omega, A]$ as $\text{Tr}\{\tilde{\omega}, \omega\}A$, we have

$$Z\langle \mathbf{Q} \rangle_{\hat{A}V} = \int dM\,dh\,dH\,dk\,d\omega\,d\tilde{\omega} \exp\left(\text{Tr}(ih + \{\tilde{\omega}, \omega\})A\right.$$

$$\left. + \text{Tr}(V + X + Y)B\right) \exp(-S)\mathbf{Q}, \tag{4.64}$$

with Z here given by the expression on the right-hand side of Eq. (4.64) with $\mathbf{Q}$ replaced by unity. Ward identities follow from the fact that the unrestricted measure dM is invariant under a shift of any matrix M_d by a constant δM_d, which under the assumption that surface terms related to the shift vanish, implies

$$0 = \int dM\,dh\,dH\,dk\,d\omega\,d\tilde{\omega}\,\delta M_d \left(\exp\left(\text{Tr}(ih + \{\tilde{\omega}, \omega\})A\right.\right.$$

$$\left.\left. + \text{Tr}(V + X + Y)B\right) \exp(-S)\mathbf{Q}\right). \tag{4.65a}$$

When S and $\mathbf{Q}$ are varied with respect to M_d, the factor δM_d can be cyclically permuted as usual to the right in each term of the varied trace polynomials, giving the familiar formulas

$$\delta_{M_d}S = \text{Tr}\frac{\delta S}{\delta M_d}\delta M_d,$$

$$\delta_{M_d}\mathbf{Q} = \text{Tr}\frac{\delta \mathbf{Q}}{\delta M_d}\delta M_d. \tag{4.65b}$$

Carrying through the variations of all terms of Eq. (4.65a), and dividing by Z, we are left with an expression of the form

$$0 = \text{Tr}\langle W_d \rangle_{\hat{A}V}\delta M_d. \tag{4.66a}$$

However, since δM_d is an arbitrary self-adjoint matrix, the vanishing of the real and imaginary parts of Eq. (4.66a) implies the matrix identity

$$0 = \langle W_d \rangle_{\hat{A}V}. \tag{4.66b}$$

For $d = 3, \ldots, D$, the variation δ_{M_d} in Eq. (4.65a) acts only on the product $\exp(-S)\mathbf{Q}$, and we have

$$W_d = \frac{\delta \mathbf{Q}}{\delta M_d} - \mathbf{Q}\frac{\delta S}{\delta M_d}. \tag{4.67a}$$

However, for $d = 1$ and $d = 2$, corresponding to $M_1 = A$ and $M_2 = B$, there are additional contributions to the Ward identities arising from variations of the traces involving A and B in the first exponential on the right-hand side of Eq. (4.65a), which arose from the unitary fixing procedure. Explicitly, we have

$$W_1 = (ih + \{\tilde{\omega}, \omega\})Q + \frac{\delta Q}{\delta A} - Q\frac{\delta S}{\delta A},$$

$$W_2 = (V + X + Y)Q + \frac{\delta Q}{\delta B} - Q\frac{\delta S}{\delta B}. \tag{4.67b}$$

Hence from Eq. (4.63) we are able to get explicit forms of all of the Ward identities, including those obtained by varying the matrices singled out in the unitary invariance fixing. Note that were we to employ the original ensemble average
of Eq. (4.51a), which has no unitary fixing, in deriving the Ward identities, then
Eq. (4.52b) implies that we would only obtain the trace of the matrix relation of
Eq. (4.66b). In other words, unitary fixing is essential for extracting the full content
of the Ward identities; without it, all nontrivial matrix structure is averaged out.

5

The emergence of quantum field dynamics

In Chapter 2, we have seen that a generic feature of matrix models with a global unitary invariance is the existence of a conserved operator $\tilde{C}$, which is given in Eq. (2.6) as the sum of bosonic q_r, p_r commutators minus the corresponding sum of fermionic anticommutators

$$\tilde{C} \equiv \sum_{r \in B} [q_r, p_r] - \sum_{r \in F} \{q_r, p_r\} = \sum_{r,u} x_r \omega_{ru} x_u, \tag{5.1}$$

and which is anti-self-adjoint when we adopt the fermion adjointness assignment of Eqs. (2.4b) and (2.18d). The operator $\tilde{C}$ and the trace quantities **H** and **N** play a role in the equilibrium statistical mechanics of trace dynamics closely analogous to that played by the energy in classical statistical physics. This analogy suggests the idea (Adler and Millard, 1996) that the canonical commutation relations of quantum field theory may arise from a trace dynamics analog of the classical theorem of equipartition of energy, in which the operator $\tilde{C}$ is effectively equipartitioned. Because $\tilde{C}$ has the dimension of action, its associated Lagrange multiplier $\tilde{\lambda}$ has the dimension of inverse action, and is the ensemble parameter that governs the magnitude of the ensemble average of $\tilde{C}$, which we shall find plays the role of the emergent Planck constant $\hbar$. If we assume that the Lagrange multiplier τ for **H** is of order the inverse Planck energy, there can be a low energy regime in which the equilibrium distribution is effectively decoupled from **H**, and is instead dominated by the $\tilde{C}$ term in the canonical ensemble. In this regime, the trace Hamiltonian still serves as generator of the effective equations of motion. If the underlying dynamics has a Poincaré invariant trace action, the Poincaré covariance of the equations of motion together with the Poincaré invariance of $\tilde{C}$ will then give rise, when the dynamics is averaged over the canonical ensemble, to a Poincaré covariant effective dynamics, with the equipartition of $\tilde{C}$ giving this effective dynamics the standard canonical commutation/anticommutation relations of quantum field theory. Thus, the combined effect of a decoupling of the effective, ensemble averaged, dynamics from the non-covariant **H** term in the canonical ensemble, and of the equipartition of $\tilde{C}$, is the emergence of relativistic quantum field theory as the low energy effective approximation to a relativistic trace dynamics.

To motivate the methods that we will use to develop these ideas, let us begin by reviewing a simple derivation (Mohling, 1982) of the classical equipartition theorem. Let $H(\{x_r\})$ be the classical Hamiltonian as a function of classical phase space variables $\{x_r\}$, and let $d\mu(\{x_r\})$ be the classical phase space integration measure. We consider the integral

$$\int d\mu \frac{\partial[x_r \exp(-\beta H)]}{\partial x_s}$$

$$= \int d\mu \delta_{rs} \exp(-\beta H)$$

$$- \int d\mu x_r \frac{\partial[\beta H]}{\partial x_s} \exp(-\beta H), \tag{5.2a}$$

the left-hand side of which is the integral of a total derivative of a function and vanishes when the function vanishes sufficiently rapidly at infinity. Assuming this, we get

$$\delta_{rs} = \frac{\int d\mu x_r \beta(\partial H/\partial x_s) \exp(-\beta H)}{\int d\mu \exp(-\beta H)}, \tag{5.2b}$$

which is the classical theorem of equipartition of energy. The method of derivation is similar to that used to derive Ward identities from functional integrals in quantum field theory (see, e.g., Kaku, 1993), and the equipartition theorem can be viewed as a Ward identity application in classical statistical mechanics.

Emulating this derivation from classical statistical physics, in Section 5.1 we derive the general Ward identity for the canonical ensemble in trace dynamics, in the presence of external sources. In Section 5.2 we show that by suitable variation of the source terms, we can get similar Ward identities obeyed by general polynomials in the effective projections $x_{r\text{eff}}$ of the matrix variables that were introduced in Eqs. (4.41b,d). In Section 5.3 we introduce the assumptions and approximations that, applied to the Ward identities of Sections 5.1 and 5.2, lead to the emergence at low energies of an effective quantum field dynamics. In Section 5.4 we take a critical look at this analysis, based on the derivation of further Ward identities, which we show lead to significant restrictions on the class of underlying theories that obey the assumptions of Section 5.3. One of these restrictions requires an approximate balance in the underlying dynamics between the numbers of bosonic and fermionic degrees of freedom, which is suggestive of supersymmetry. The derivation of Section 5.3 leads directly to an emergent Heisenberg dynamics; in Section 5.5 we derive from this the emergent Schrödinger dynamics. Finally, in

Section 5.6 we show how the emergence of quantum field dynamics from trace dynamics evades the Kochen–Specker theorem and Bell inequality arguments against a "hidden variable" completion of quantum mechanics.

5.1 The general Ward identity

We now apply a procedure, similar to that of the derivation of the classical equipartition theorem, to the statistical mechanics of matrix models. We begin by specifying some notation. Let $d\hat{\mu}$ be the measure introduced in Section 4.5, in which the integration over *one* canonical pair of dynamical variables x_R, x_{R+1} is restricted so as to break the subgroup of the global unitary group that commutes with i_{eff}. Also, let $\langle \mathcal{O} \rangle_{\widehat{\text{AV}}}$ be the average over the canonical ensemble, using this restricted measure, of a general operator depending on the $\{x_r\}$. This average is given by

$$\langle \mathcal{O} \rangle_{\widehat{\text{AV}}} = \int d\hat{\mu}\rho\mathcal{O}, \tag{5.3a}$$

with the normalized equilibrium distribution ρ given by Eq. (4.13c), as modified by the replacement of $d\mu$ by the restricted integration measure $d\hat{\mu}$ in the partition function Z. Since we shall wish to include sources, let $\langle \mathcal{O} \rangle_{\widehat{\text{AV}},j}$ be the corresponding average in the presence of a complete set $\{j_r\}$ of external sources, given by

$$\langle \mathcal{O} \rangle_{\widehat{\text{AV}},j} = \int d\hat{\mu}\rho_j\mathcal{O}, \tag{5.3b}$$

with ρ_j and the associated partition function Z_j given by the expressions of Eq. (4.15a) that include sources, now with restricted integration measure

$$\rho_j = Z_j^{-1} \exp\left(-\text{Tr}\tilde{\lambda}\tilde{C} - \tau\mathbf{H} - \eta\mathbf{N} - \sum_r \text{Tr}\,j_r x_r\right),$$

$$Z_j = \int d\hat{\mu} \exp\left(-\text{Tr}\tilde{\lambda}\tilde{C} - \tau\mathbf{H} - \eta\mathbf{N} - \sum_r \text{Tr}\,j_r x_r\right). \tag{5.3c}$$

Thus, in this notation, we have

$$\langle \mathcal{O} \rangle_{\widehat{\text{AV}}} = \langle \mathcal{O} \rangle_{\widehat{\text{AV}},0}, \tag{5.3d}$$

with the subscript 0 on the right-hand side denoting the average in which all sources are zero, that is, with $\{j_r\} = \{0\}$. In deriving the Ward identity, we shall employ the fact that the integration measure $d\mu$ is invariant under a constant shift $x_r \to x_r + \delta x_r$ of any of the dynamical variables x_r. When we use the restricted integration measure $d\hat{\mu}$, this invariance still holds, provided r is not equal to the indices R, $R+1$ of the variables x_R, x_{R+1} for which the phase space integration is restricted. In other words, with respect to the restricted measure we have

$$0 = \int d\hat{\mu}\, \delta_{x_r}(\rho\mathcal{O}), \quad r \neq R, \ R+1, \tag{5.4a}$$

with

$$\delta_{x_r}\mathcal{A} = \mathcal{A}|_{x_r + \delta x_r} - \mathcal{A}|_{x_r}. \tag{5.4b}$$

With this notation established, we begin our derivation by considering

$$Z_j \langle \mathrm{Tr}\{\tilde{C}, i_{\mathrm{eff}}\} W \rangle_{\hat{\mathrm{AV}},j} = \int d\hat{\mu}\, \exp\left(-\mathrm{Tr}\tilde{\lambda}\tilde{C} - \tau\mathbf{H} - \eta\mathbf{N}\right.$$

$$\left. - \sum_r \mathrm{Tr}\, j_r x_r \right) \mathrm{Tr}\{\tilde{C}, i_{\mathrm{eff}}\} W, \tag{5.5a}$$

with W any bosonic (that is, even grade) polynomial function of the dynamical variables, and where the trace Tr is understood to act on the product of all factors standing to its right. (Later on, we will successively let W be either the operator Hamiltonian H from which the trace Hamiltonian $\mathbf{H}$ is constructed, a single dynamical variable x_t, or an operator G from which a trace canonical generator $\mathbf{G}$ is constructed, in order to derive respectively in the emergent quantum theory the Heisenberg equation of motion, the canonical algebra, or the form of a general unitary canonical transformation.) Using Eqs. (5.4a,b), when we make a shift of x_s, $s \neq R$, $R+1$ in the integrand of Eq. (5.5a), we have

$$0 = \int d\hat{\mu}\, \delta_{x_s} \left[\exp\left(-\mathrm{Tr}\tilde{\lambda}\tilde{C} - \tau\mathbf{H} - \eta\mathbf{N} - \sum_r \mathrm{Tr}\, j_r x_r\right) \mathrm{Tr}\{\tilde{C}, i_{\mathrm{eff}}\} W \right], \tag{5.5b}$$

which on applying the chain rule for differentiation becomes

$$0 = \int d\hat{\mu}\, \exp\left(-\mathrm{Tr}\tilde{\lambda}\tilde{C} - \tau\mathbf{H} - \eta\mathbf{N} - \sum_r \mathrm{Tr}\, j_r x_r\right)$$

$$\times \left[(-\mathrm{Tr}\tilde{\lambda}\delta_{x_s}\tilde{C} - \tau\delta_{x_s}\mathbf{H} - \eta\delta_{x_s}\mathbf{N} - \mathrm{Tr}\, j_s\delta x_s)\mathrm{Tr}\{\tilde{C}, i_{\mathrm{eff}}\} W + \delta_{x_s}\mathrm{Tr}\{\tilde{C}, i_{\mathrm{eff}}\} W \right]. \tag{5.5c}$$

We now have to evaluate the variations with respect to x_s appearing in the various terms on the right-hand side of Eq. (5.5c). From Eq. (5.1) for $\tilde{C}$, we have

$$\mathrm{Tr}\tilde{\lambda}\delta_{x_s}\tilde{C} = \mathrm{Tr}\tilde{\lambda} \sum_r (\delta x_s \omega_{sr} x_r + x_r \omega_{rs} \delta x_s), \tag{5.6a}$$

which using $\omega_{sr} = -\epsilon_r \omega_{rs}$ (cf. Eq. (1.17)) can be rewritten as

$$\text{Tr}\tilde{\lambda}\delta_{x_s}\tilde{C} = \text{Tr}\tilde{\lambda}\sum_r \omega_{rs}(-\epsilon_r \delta x_s x_r + x_r \delta x_s). \tag{5.6b}$$

Cyclically permuting x_r to the left in the first term using Eqs. (1.1a,b), this simplifies to a commutator

$$\text{Tr}\tilde{\lambda}\delta_{x_s}\tilde{C} = \text{Tr}\left[\tilde{\lambda}, \sum_r \omega_{rs}x_r\right]\delta x_s. \tag{5.6c}$$

We next consider $\delta_{x_s}\mathbf{H}$, which by Eqs. (1.3b) and (1.15b) through (1.17) is given by

$$\delta_{x_s}\mathbf{H} = \text{Tr}\frac{\delta\mathbf{H}}{\delta x_s}\delta x_s = \sum_r \omega_{rs}\text{Tr}\dot{x}_r \delta x_s. \tag{5.7}$$

Note that here, and in the subsequent formulas, $\dot{x}_r dt$ represents the forward time step of the variable x_r, considered as a function of the dynamical variables $\{x_s\}$ on the time slice on which averages over the canonical ensemble are evaluated. Turning next to the evaluation of $\delta_{x_s}\mathbf{N}$, let us define $\tilde{\omega}_{rs}$ by

$$\tilde{\omega}_{rs} = \text{diag}(0, \ldots, 0, \Omega_B, \ldots, \Omega_B), \tag{5.8a}$$

where the skew symmetric 2×2 matrix Ω_B is defined in Eq. (1.16b). Recalling our convention that we list all bosonic variables before all fermionic ones in the $2D$-dimensional phase space vector x_r, the definition of Eq. (5.8a) states that $\tilde{\omega}_{rs}$ acts as 0 on any bosonic phase space pair q_r, p_r and acts as Ω_B on any fermionic phase space pair q_r, p_r, so that

$$\sum_{r,u} \tilde{\omega}_{ru}x_r x_u = \sum_{r\in F}[q_r, p_r], \tag{5.8b}$$

giving the quantity appearing in Eq. (2.2a) defining $\mathbf{N}$. Using the index interchange relation $\tilde{\omega}_{sr} = \epsilon_r \tilde{\omega}_{rs}$, we then find

$$\delta_{x_s}\mathbf{N} = \frac{1}{2}i\sum_r \text{Tr}(\tilde{\omega}_{sr}\delta x_s x_r + \tilde{\omega}_{rs}x_r \delta x_s) = i\sum_r \tilde{\omega}_{rs}\text{Tr}x_r \delta x_s. \tag{5.8c}$$

This completes the calculation of variations of terms that come from the exponent in the canonical ensemble ρ.

The remaining terms come from

$$\delta_{x_s}\text{Tr}\{\tilde{C}, i_{\text{eff}}\}W = \text{Tr}(\{\delta_{x_s}\tilde{C}, i_{\text{eff}}\}W + \{\tilde{C}, i_{\text{eff}}\}\delta_{x_s}W)$$
$$= \text{Tr}(\{i_{\text{eff}}, W\}\delta_{x_s}\tilde{C} + \{\tilde{C}, i_{\text{eff}}\}\delta_{x_s}W), \tag{5.9a}$$

where to obtain the second line we have used the cyclic identity of Eq. (1.2). For the first term on the right-hand side of Eq. (5.9a), we use Eq. (5.6c), with $\tilde{\lambda}$ replaced by $\{i_{\text{eff}}, W\} = 2i_{\text{eff}}W_{\text{eff}}$ (where we have used Eq. (4.41d)), giving

$$\text{Tr}\{i_{\text{eff}}, W\}\delta_{x_s}\tilde{C} = \text{Tr}\left[2i_{\text{eff}}W_{\text{eff}}, \sum_r \omega_{rs}x_r\right]\delta x_s. \tag{5.9b}$$

To evaluate the second term on the right-hand side of Eq. (5.9a), we write the operator structure of $\delta_{x_s}W$ in the form

$$\delta_{x_s}W = \sum_\ell W_s^{L\ell}\delta x_s W_s^{R\ell}, \tag{5.10a}$$

where ℓ is a composite index that labels each monomial in the polynomial W, as well as each occurrence of x_s in the respective monomial term. In this notation we have

$$\frac{\delta W}{\delta x_s} = \sum_\ell \epsilon_\ell W_s^{R\ell}W_s^{L\ell}, \tag{5.10b}$$

with ϵ_ℓ the grading factor appropriate to $W_s^{R\ell}$ and to $W_s^{L\ell}x_s$ (which must both be of the same grade since we have defined W to be bosonic). From the definition of Eq. (5.10a), we obtain

$$\text{Tr}\{\tilde{C}, i_{\text{eff}}\}\delta_{x_s}W = \sum_\ell \epsilon_\ell \text{Tr}W_s^{R\ell}\{\tilde{C}, i_{\text{eff}}\}W_s^{L\ell}\delta x_s. \tag{5.10c}$$

When these results are collected and substituted back into Eq. (5.5c), this equation takes the form (after multiplication by Z_j^{-1})

$$0 = \langle\text{Tr}\Sigma_s\delta x_s\rangle_{\hat{\text{AV}},j}, \tag{5.11a}$$

with Σ_s a shorthand for the sum of the contributions coming from Eqs. (5.6a) through (5.10c)

$$\Sigma_s = \left(-\left[\tilde{\lambda}, \sum_r \omega_{rs}x_r\right] - \tau\sum_r \omega_{rs}\dot{x}_r - i\eta\sum_r \tilde{\omega}_{rs}x_r - j_s\right)2\text{Tr}\tilde{C}i_{\text{eff}}W_{\text{eff}}$$

$$+ 2\left[i_{\text{eff}}W_{\text{eff}}, \sum_r \omega_{rs}x_r\right] + \sum_\ell \epsilon_\ell W_s^{R\ell}\{\tilde{C}, i_{\text{eff}}\}W_s^{L\ell}. \tag{5.11b}$$

Since the variation δx_s is arbitrary, subject to the adjointness restrictions on x_s, and since both the real and imaginary parts of Eq. (5.11a) must vanish separately, we can conclude from Eq. (5.11a) that

$$\langle\Sigma_s\rangle_{\hat{\text{AV}},j} = 0. \tag{5.11c}$$

To get the final form of the Ward identity, we perform several algebraic manipulations on Eq. (5.11c). First of all, we form the effective projection by taking one half of the anticommutator of Eq. (5.11c) with i_{eff} (cf. Eq.(4.41d)). Since $\tilde{\lambda} = \lambda i_{\text{eff}}$ (cf. Eq. (4.39b)), and since i_{eff} commutes with $x_{r\text{eff}}$, the first term on the right-hand side of Eq. (5.11b) drops out (in other words, only terms that anticommute with i_{eff} remain in the commutator $[\tilde{\lambda}, x_r]$) and we are left with

$$\langle \Sigma_{s\text{eff}} \rangle_{\hat{\text{AV}}, j} = 0, \tag{5.12a}$$

with $\Sigma_{s\text{eff}}$ given by

$$\Sigma_{s\text{eff}} = \left(-\tau \sum_r \omega_{rs} \dot{x}_{r\text{eff}} - i\eta \sum_r \tilde{\omega}_{rs} x_{r\text{eff}} - j_{s\text{eff}} \right) 2\text{Tr}\tilde{C} i_{\text{eff}} W_{\text{eff}}$$

$$+ 2 \left[i_{\text{eff}} W_{\text{eff}}, \sum_r \omega_{rs} x_{r\text{eff}} \right] + \sum_\ell \epsilon_\ell (W_s^{R\ell} \{\tilde{C}, i_{\text{eff}}\} W_s^{L\ell})_{\text{eff}}. \tag{5.12b}$$

(This step of the derivation requires use of the canonical ensemble: in the microcanonical ensemble, the analogous term arising from the $\tilde{C}$ dependence of the ensemble Γ of Eq. (4.18a) does not have the form of a commutator $[\tilde{C}^{\text{micro}}, \ldots]$, where $\tilde{C}^{\text{micro}}$ is the sharp value of $\tilde{C}$ in the microcanonical ensemble, and so does not vanish on taking a suitably defined effective projection. Put another way, if one represents the delta function $\delta(\tilde{C} - \tilde{C}^{\text{micro}})$ appearing in the microcanonical ensemble as an integral

$$\delta(\tilde{C} - \tilde{C}^{\text{micro}}) \propto \int d\tilde{\lambda} \exp\left(i\text{Tr}\tilde{\lambda}(\tilde{C} - \tilde{C}^{\text{micro}})\right), \tag{5.12c}$$

then the parameter $\tilde{\lambda}$ appearing in a commutator structure analogous to the first term in Eq. (5.11b) is an integration variable, and so an effective projection using a fixed i_{eff} does not eliminate this term.)

Next, we multiply Eqs. (5.12a,b) by $\frac{1}{2}\omega_{us}$, where u does not lie in the canonical pair containing the indices R, $R+1$ of variables x_R, x_{R+1} for which the phase space integration is restricted. We then sum s over index values within the canonical pair containing u using Eq. (1.17), which gives

$$\sum_s \omega_{us} \omega_{rs} = \delta_{ur}, \tag{5.13a}$$

and employ the analogous formula involving $\tilde{\omega}_{rs}$

$$\sum_s \omega_{us} \tilde{\omega}_{rs} = -\xi_u \delta_{ur}, \tag{5.13b}$$

with $\xi_u = 0$ for any bosonic x_u and with $\xi_u = 1\ (-1)$ for x_u a fermionic $q\ (p)$. This gives the final result

$$\langle \Lambda_{u\text{eff}} \rangle_{\hat{\text{AV}},j} = 0, \tag{5.14a}$$

with $\Lambda_{u\text{eff}}$ given by

$$\Lambda_{u\text{eff}} = \frac{1}{2} \sum_s \omega_{us} \Sigma_{s\text{eff}}$$

$$= \left(-\tau \dot{x}_{u\text{eff}} + i\eta \xi_u x_{u\text{eff}} - \sum_s \omega_{us} j_{s\text{eff}} \right) \text{Tr} \tilde{C} i_{\text{eff}} W_{\text{eff}} \tag{5.14b}$$

$$+ [i_{\text{eff}} W_{\text{eff}}, x_{u\text{eff}}] + \sum_{s,\ell} \omega_{us} \epsilon_\ell \left(W_s^{R\ell} \frac{1}{2} \{\tilde{C}, i_{\text{eff}}\} W_s^{L\ell} \right)_{\text{eff}}.$$

This equation holds for all index values u not lying in the canonical pair containing the indices $R,\ R+1$; henceforth this restriction will be implicit, and will usually not be restated explicitly each time the Ward identity formula of Eqs. (5.14a,b) is used (and similarly for its analogs derived in Section 5.4). The additional terms that are present in the Ward identities in which u belongs to the pair $R,\ R+1$, arising from variation of the unitary fixing conditions on the integration measure, can be calculated (at least formally) by the method given in Section 4.6.

5.2 Variation of the source terms

The Ward identity of Eqs. (5.14a,b) is exact, and still includes the full source term structure. Our next step is to show how, through variation of the source terms, we can generate similar Ward identities involving general polynomials in the effective projections $x_{r\text{eff}}$ of the dynamical variables. To see how such polynomials can be generated, we use the notation introduced in Eq. (4.41b) to rewrite the source term in the canonical ensemble by making the decompositions $x_r = x_{r\text{eff}} + x_{r12}$ and $j_r = j_{r\text{eff}} + j_{r12}$. Since for arbitrary operators A and B we have $\text{Tr} A_{\text{eff}} B_{12} = \text{Tr}(-\frac{1}{2} i_{\text{eff}})\{i_{\text{eff}}, A_{\text{eff}}\} B_{12} = \text{Tr}(-\frac{1}{2} i_{\text{eff}}) A_{\text{eff}} \{i_{\text{eff}}, B_{12}\} = 0$, we get $\text{Tr} j_r x_r = \text{Tr}(j_{r\text{eff}} x_{r\text{eff}} + j_{r12} x_{r12})$, and so varying Z_j with respect to $j_{r\text{eff}}$ brings down a factor of $\text{Tr} \delta j_{r\text{eff}} x_{r\text{eff}}$. Stripping away the general variation $\delta j_{r\text{eff}}$ from the left then leaves us with a matrix factor $x_{r\text{eff}}$. (We choose not to strip away the variation $\delta j_{r\text{eff}}$ from the right, because by the cyclic identities this would leave an extra factor ϵ_r to be kept track of.) As we shall see, doing this repeatedly allows one to build up general polynomials, i.e., linear combinations with c-number coefficients of monomials of the form $x_{r\text{eff}} x_{s\text{eff}} \cdots$.

We can now state the result to be demonstrated in this section as follows: defining $\mathcal{D}x_{u\text{eff}}$ by

$$\mathcal{D}x_{u\text{eff}} = (-\tau\dot{x}_{u\text{eff}} + i\eta\xi_u x_{u\text{eff}})\text{Tr}\tilde{C}i_{\text{eff}}W_{\text{eff}}$$

$$+ [i_{\text{eff}}W_{\text{eff}}, x_{u\text{eff}}] + \sum_{s,\ell}\omega_{us}\epsilon_\ell \left(W_s^{R\ell}\frac{1}{2}\{\tilde{C}, i_{\text{eff}}\}W_s^{L\ell}\right)_{\text{eff}}, \qquad (5.15a)$$

we can rewrite Eq. (5.14a) as

$$\langle \mathcal{D}x_{u\text{eff}}\rangle_{\hat{\text{AV}},j} - \sum_s \omega_{us} j_{s\text{eff}}\langle\text{Tr}\tilde{C}i_{\text{eff}}W_{\text{eff}}\rangle_{\hat{\text{AV}},j} = 0. \qquad (5.15b)$$

Then we shall show that Eq. (5.15b) also implies the relations at zero sources

$$\langle S_L(x_{t\text{eff}})(\mathcal{D}S(x_{r\text{eff}}))S_R(x_{t\text{eff}})\rangle_{\hat{\text{AV}},0} = 0. \qquad (5.15c)$$

Here S is a polynomial (with c-number coefficients) in the effective variables $x_{r\text{eff}}$ (i.e., $x_{r\text{eff}}, x_{s\text{eff}}, \ldots$ which we call generically $x_{r\text{eff}}$), with the functional form of S subject only to the restriction that it should not depend on the variables in the canonical pair containing the indices R, $R+1$. In order to be able to go "off-shell" when we make contact later on with quantum field theory, we have included in Eq. (5.15c) left and right polynomials S_L and S_R (also with c-number coefficients), with the property that for all x_t in $S_{L,R}$ and all x_r in S, the structure constant ω_{tr} is zero. In other words, for the time being we assume that $S_{L,R}$ do not contain variables that are the canonical conjugates of those that appear in S. A sufficient condition for this restriction to be dropped is discussed at the end of this section. The action of $\mathcal{D}$ on S in Eq. (5.15c) is defined by the Leibniz product rule

$$\mathcal{D}(x_{r\text{eff}}x_{s\text{eff}}) = (\mathcal{D}x_{r\text{eff}})x_{s\text{eff}} + x_{r\text{eff}}(\mathcal{D}x_{s\text{eff}}), \qquad (5.15d)$$

and its extension to products of any number of factors $x_{r\text{eff}}$ with general index values.

Since the case of polynomial S, S_L, and S_R with c-number coefficients can be obtained by linearity from the case of monomial S, S_L, and S_R, it suffices to prove Eq. (5.15c) for the case in which we are dealing with monomials. Now for a $\mathcal{D}$ defined to act on products by the Leibniz product rule, and for a monomial S, we have

$$\mathcal{D}S(x_{r\text{eff}}) = \sum_v S(x_{r\text{eff}}, r \neq v; \mathcal{D}x_{v\text{eff}}), \qquad (5.16a)$$

with v indexing each occurrence of each $x_{r\text{eff}}$ in S, and with all factor orderings left undisturbed by the action of $\mathcal{D}$. In other words, leaving factor orderings unchanged, we include on the right-hand side of Eq. (5.16a) a term with $\mathcal{D}$ acting on each individual variable $x_{r\text{eff}}$ in S. Hence to establish Eq. (5.15c), it suffices to prove that for monomials S, S_L, and S_R, we have

$$\langle S_L(x_{t\text{eff}}) \sum_v S(x_{r\text{eff}}, r \neq v; \mathcal{D}x_{v\text{eff}}) S_R(x_{t\text{eff}}) \rangle_{\hat{\text{AV}},0} = 0. \tag{5.16b}$$

To prove this assertion, we begin by multiplying Eq. (5.15b) by $Z_j \delta j_{u\text{eff}}$ and taking the trace, giving

$$Z_j \langle \text{Tr} \delta j_{u\text{eff}} \mathcal{D}x_{u\text{eff}} \rangle_{\hat{\text{AV}},j} - Z_j \sum_s \omega_{us} \text{Tr}(\delta j_{u\text{eff}} j_{s\text{eff}}) \langle \text{Tr} \tilde{C}i_{\text{eff}} W_{\text{eff}} \rangle_{\hat{\text{AV}},j} = 0. \tag{5.17a}$$

We now make sequential, independent variations of the sources $j_{r\text{eff}}$ associated with all $x_{r\text{eff}}$ in the monomial S other than the one $x_{u\text{eff}}$ that is explicitly exhibited in Eq. (5.17a), as well as variations of the sources associated with all $x_{t\text{eff}}$ in the monomials $S_{L,R}$. To get repeated factors of some particular $x_{T\text{eff}}$, we make independent source variations for each, and after all variations have been performed, we set all sources equal to zero. In the second term in Eq. (5.17a), if $j_{s\text{eff}}$ is not varied in this process, it makes a vanishing contribution after the sources are set to zero. The case when $j_{s\text{eff}}$ is varied can only arise when we are varying with respect to the source for a variable $x_{v\text{eff}}$ contained in S, since by hypothesis the coefficient tensor ω_{ut} vanishes for all $x_{t\text{eff}}$ in $S_{L,R}$. In general, if we focus on two variables $x_{v\text{eff}}$ and $x_{u\text{eff}}$ for which $\omega_{uv} \neq 0$, the monomial S will have the form $\ldots x_{u\text{eff}} \ldots x_{v\text{eff}} \ldots$ (or a similar expression with the roles of u and v interchanged; we do not assume symmetrization of this structure over u and v), with $\ldots$ denoting factors that are not explicitly exhibited. Correspondingly, the derivative $\mathcal{D}S$ formed by use of the Leibniz rule will have the form

$$\ldots (\mathcal{D}x_{u\text{eff}}) \ldots x_{v\text{eff}} \ldots + \ldots x_{u\text{eff}} \ldots (\mathcal{D}x_{v\text{eff}}) \ldots. \tag{5.17b}$$

In forming the first term in Eq. (5.17b) by source variations, the second term of Eq. (5.17a) will contribute the second variation expression

$$\ldots \omega_{uv} \text{Tr}(\delta j_{u\text{eff}} \delta j_{v\text{eff}}) \ldots, \tag{5.17c}$$

while in forming the second term in Eq. (5.17b), there will be a corresponding second variation expression

$$\ldots \omega_{vu} \text{Tr}(\delta j_{v\text{eff}} \delta j_{u\text{eff}}) \ldots. \tag{5.17d}$$

Since both second variation expressions multiply identical factors, they contribute through their sum

$$\ldots [\omega_{uv} \text{Tr}(\delta j_{u\text{eff}} \delta j_{v\text{eff}}) + \omega_{vu} \text{Tr}(\delta j_{v\text{eff}} \delta j_{u\text{eff}})] \ldots. \tag{5.17e}$$

However, this expression vanishes, because in the bosonic (fermionic) case ω_{uv} is antisymmetric (symmetric) in u, v, while the trace multiplying it is symmetric (antisymmetric) in u, v.

Thus, in forming Eq. (5.16b) by source variation, the explicit source term in Eq. (5.15b) does not contribute after all sources are set equal to zero, and so the effect of multiple source variations is to lead to an expression of the form

$$Z\langle \prod_{r \neq u} \mathrm{Tr}(\delta j_{r\mathrm{eff}} x_{r\mathrm{eff}}) \mathrm{Tr}(\delta j_{u\mathrm{eff}} \mathcal{D} x_{u\mathrm{eff}}) \rangle_{\hat{\mathrm{AV}},0} = 0, \qquad (5.17f)$$

where r indexes variables that appear in both S and $S_{L,R}$. Since the source variations in Eq. (5.17f) are all independent, we can use them to "break open" the traces as in the discussion of Eqs. (1.4a–d), giving the expression

$$Z\langle \prod_{r \neq u} (x_{r\mathrm{eff}})_{n_r m_r} (\mathcal{D} x_{u\mathrm{eff}})_{n_u m_u} \rangle_{\hat{\mathrm{AV}},0} = 0, \qquad (5.17g)$$

with the average now involving a product of matrix elements. By linking the matrix elements in Eq. (5.17g) in the appropriate order, we can form the general matrix element of $\mathcal{D}$ acting on a general monomial S, multiplied on the left and right by monomials $S_{L,R}$ containing variables $x_{t\mathrm{eff}}$ that are not linked by ω_{rt} to any variable $x_{r\mathrm{eff}}$ in S. This then gives the identity of Eq. (5.16b), completing the proof of Eq. (5.15c).

Up to this point we have made no special assumptions about the nature of the bosonic polynomial W appearing in the Ward identity. As we shall see later, in the cases of greatest interest for applications of the Ward identity, the trace $\mathrm{Tr}\tilde{C}_{i\mathrm{eff}} W_{\mathrm{eff}}$ can be approximated as zero inside canonical averages in which the source currents j have been set to zero. In this case, the derivation just given simplifies considerably, since there are then no second variation terms arising from varying $j_{s\mathrm{eff}}$ in Eq. (5.17a), because such second variation terms all appear with coefficient zero when the sources have been turned off. The symmetrization argument of Eqs. (5.17b–e) can then be dispensed with, and, more significantly, no restriction on the arguments x_t of the left and right polynomials $S_{L,R}$ is then needed. Thus for these special choices of W the identity of Eq. (5.15c) holds for completely general polynomials $S_L(x_{t\mathrm{eff}})$ and $S_R(x_{t\mathrm{eff}})$, giving a completely "off-shell" version of the basic Ward identity which asserts the vanishing of $\mathcal{D}S(x_{r\mathrm{eff}})$. Once the Ward identity has been extended off-shell in this way, we are assured that fluctuations of the effective variables $x_{r\mathrm{eff}}$ around their canonical ensemble average values have been taken into account, and do not invalidate the argument for the emergence of a quantum theory structure given in the next section.

5.3 Approximations/assumptions leading to the emergence
of quantum theory

Starting from Eqs. (5.15a–c), we proceed now to show that, with a few plausible assumptions and approximations, the general formalism of quantum field theory emerges. The assumptions and approximations that we make all have the effect of simplifying the structure of $\mathcal{D}x_{u\text{eff}}$. They are:

(1)   We assume that the support properties of $\dot{x}_{u\text{eff}}$ and of $\tilde{C}_{\text{eff}} = -\frac{1}{2}i_{\text{eff}}\{i_{\text{eff}}, \tilde{C}\}$, as functions defined on the operator phase space, are such that the term

$$- \tau \dot{x}_{u\text{eff}} \operatorname{Tr} \tilde{C} i_{\text{eff}} W_{\text{eff}}$$
$$= -\tau \dot{x}_{u\text{eff}} \operatorname{Tr} \tilde{C}_{\text{eff}} i_{\text{eff}} W_{\text{eff}} \qquad (5.18a)$$

in Eq. (5.15a) can be neglected. Specifically, we shall assume that this comes about in the following way. We assume that each dynamical variable $x_{u\text{eff}}$ can be split into "fast" and "slow" parts or components, so that we can write $x_{u\text{eff}} = x_{u\text{eff}}^{\text{fast}} + x_{u\text{eff}}^{\text{slow}}$. We identify the time scale τ and mass τ^{-1} with the "fast" or "high" physical scale given by the Planck scale, and we assume that the underlying theory develops a mass hierarchy, so that observed physics corresponds to "slow" components $x_{u\text{eff}}^{\text{slow}}$ that are very slowly varying in comparison to the time τ. The "fast" parts of the dynamical variables will be insensitive to large-scale boundary conditions, such as experimental probes, whereas the "slow" parts are sensitive to such boundary conditions, and are what we observe in our experiments. We further assume that the "fast" components $x_{u\text{eff}}^{\text{fast}}$ have disjoint support on the operator phase space from the support of $\tilde{C}_{\text{eff}}$, so that averages of products of the fast components with $\tilde{C}_{\text{eff}}$ are suppressed. Then for the "slow" components $x_{u\text{eff}}^{\text{slow}}$, the contribution to Eq. (5.18a) will be small because it is suppressed by one power of the mass hierarchy, while for the "fast" components $x_{u\text{eff}}^{\text{fast}}$, for which $\dot{x}_{u\text{eff}}^{\text{fast}}$ may not be negligibly small, the contribution to Eq. (5.18a) will be small by virtue of the assumed support properties. A detailed discussion of the support assumption and its implications will be given in the next section. (This assumption could perhaps be weakened to allow for the possibility that the 'fast' components of $x_{u\text{eff}}$ have regions of common support with $\tilde{C}_{\text{eff}}$, with rapid relative phase oscillations making the contribution of Eq. (5.18a) to the Ward identities very small. However, we shall not pursue this possibility further in what follows.)

(2)   We assume that the "chemical potential" η is zero or very small, so that the term

$$i\eta \xi_u x_{u\text{eff}} \operatorname{Tr} \tilde{C} i_{\text{eff}} W_{\text{eff}} \qquad (5.18b)$$

can be neglected. Since this term is identically zero for u bosonic (cf. the line following Eq. (5.13b)), this assumption is only operative in the fermionic sector of the theory. As we have seen in Section 2.5, the trace fermion number is odd under interchange of fermionic canonical coordinates and momenta, whereas $\tilde{C}$ is even, and it is easy to see from the construction of Section 4.1 that the integration measure is also even. Hence when the trace Hamiltonian $\mathbf{H}$ is even under this interchange, then taking

η to be exactly zero corresponds to choosing an ensemble in which the zero-source ensemble average of the trace fermion number **N** is zero.

(3) We assume that in the final term of Eq. (5.15a), as a leading approximation when the number of degrees of freedom is very large, we can replace the extensive conserved quantity $\tilde{C}_{\text{eff}}$ by its zero-source ensemble average $\langle \tilde{C}_{\text{eff}} \rangle_{\text{AV}}$. Moreover, as noted in Section 4.5, when there are very many degrees of freedom we assume that this unitary fixed average is the same as the unfixed average $\langle \tilde{C}_{\text{eff}} \rangle_{\text{AV}} = i_{\text{eff}} \hbar$. In other words, in the final term of Eq. (5.15a), we neglect the fluctuations of $\tilde{C}_{\text{eff}}$ over the canonical ensemble; these fluctuations will be added back in Chapter 6, where we argue that they are responsible for state vector reduction. (As discussed in the next section, the corresponding replacement of $\tilde{C}_{\text{eff}}$ by its ensemble average cannot be made in the τ term in Eq. (5.15a), since this would change the support properties in a significant way.) With this assumption, the final term of Eq. (5.15a) becomes

$$-\hbar \sum_{s\ell} \omega_{us} \epsilon_{\ell} (W_s^{R\ell} W_s^{L\ell})_{\text{eff}} = -\hbar \sum_s \omega_{us} \left(\frac{\delta \mathbf{W}}{\delta x_s} \right)_{\text{eff}}, \qquad (5.18c)$$

where in simplifying we have made use of Eq. (5.10b).

(4) We similarly assume that in the coefficient of $j_{s\text{eff}}$ in the second term in Eqs. (5.15b) and (5.17a), we can also replace $\tilde{C}_{\text{eff}}$ by $i_{\text{eff}} \hbar$ after the sources have been set to zero. Thus the unwanted second variation terms in the zero-source Ward identities, arising from variation of this term with respect to the source current, all have a coefficient of the form $\langle P(x_{r\text{eff}}) \text{Tr} W_{\text{eff}} \rangle_{\hat{A}V,0}$, with P some polynomial in its arguments arising from the repeated variations of the source terms in the canonical ensemble. Since $\text{Tr} W_{\text{eff}} = \text{Tr}(W - i_{\text{eff}} W i_{\text{eff}})/2 = \text{Tr} W = \mathbf{W}$, this coefficient can equally well be written as $\langle P(x_{r\text{eff}}) \mathbf{W} \rangle_{\hat{A}V,0}$, and vanishes for those W for which $\mathbf{W}$ is effectively zero inside zero-source averages. The reasons that we shall give for expecting $\mathbf{W}$ to be effectively zero will depend on our choice for W in the enumeration of cases that follows. We shall generally assume that when $\mathbf{W}$ is a conserved trace generator with $\langle \mathbf{W} \rangle_{\text{AV}} = 0$, then we can approximate $\mathbf{W}$ as zero inside canonical averages $\langle \mathbf{W} \rangle_{\hat{A}V,j_{\text{eff}}}$ in the presence of sources j_{eff}, so that $\langle P(x_{r\text{eff}}) \mathbf{W} \rangle_{\hat{A}V,0} \simeq 0$. (However, this assumption on $\mathbf{W}$ can be dispensed with when the condition, that ω_{tr} should vanish for all arguments x_t in $S_{L,R}$ and all arguments x_r in S, is satisfied, since then the unwanted second variation terms are absent. This happens, for example, when the polynomials S, S_L, S_R in Eq. (5.15c) are constructed solely from canonical coordinates q, with no arguments that are canonical momenta p.)

With assumptions (1)–(3), Eq. (5.15a) simplifies dramatically to take the form

$$\mathcal{D}x_{u\text{eff}} = i_{\text{eff}}[W_{\text{eff}}, x_{u\text{eff}}] - \hbar \sum_s \omega_{us} \left(\frac{\delta \mathbf{W}}{\delta x_s} \right)_{\text{eff}}, \qquad (5.19a)$$

where we have used the fact that i_{eff} commutes with all effective quantities to pull it outside the commutator in the first term. Consistent with our assumptions (1) and (3) of a decoupling of the "fast" dynamics from the low energy phenomenological theory, we shall assume that W_{eff} of Eq. (5.19a), in its action on the low energy or "slow" degrees of freedom, can be represented as a suitable operator function $W_{eff}(x_{reff})$ constructed from the x_{reff}, with the influence of the "fast" degrees of freedom appearing only through some set of renormalized couplings and through the average $i_{eff}\hbar$ of $\tilde{C}$. We can now use this simplified form of $\mathcal{D}$ in the Ward identity with sources given in Eq. (5.15b), and in the Ward identity obtained after source variations given in Eq. (5.15c). Assumption (4) allows us to use Eq. (5.15c) without restrictions on the arguments x_t of the left and right polynomial $S_{L,R}$, effectively taking Eq. (5.15c) "off-shell" and allowing us to interpret the vanishing of the right-hand side of Eq. (5.19a) as an operator identity in the weak sense in the effective theory.

As discussed in Adler and Millard (1996), specialization of the operator polynomial W gives a number of important results.

(A) First, let us take W in Eq. (5.19a) to be the operator Hamiltonian $H = \sum_r p_r \dot{q}_r - L$, with $\mathbf{H} = \mathrm{Tr}H$ the corresponding trace Hamiltonian, as in Eq. (1.9b). Then by using the trace dynamics equation of motion of Eq. (1.15b), Eq. (5.19a) simplifies to

$$\mathcal{D}x_{ueff} = i_{eff}[H_{eff}, x_{ueff}] - \hbar\dot{x}_{ueff}. \tag{5.19b}$$

Using Eq. (5.19b) in Eq. (5.15c), we learn that x_{ueff} obeys an effective Heisenberg picture equation of motion, which holds when sandwiched between polynomials $S_{L,R}$ that do not contain x_u, and averaged over the zero-source canonical ensemble. We now make the assumption that the canonical ensemble has vanishing average trace energy $\mathbf{H}$. Neglecting the fluctuations of the extensive quantity $\mathbf{H}$ over the canonical ensemble, which will be small for a large system, this also implies that $\langle P(x_{reff})\mathbf{H}\rangle_{\hat{A}V,0}$ vanishes. This allows us to neglect the unwanted source current variation terms arising from Eq. (5.17a), and so the restriction that $S_{L,R}$ not contain x_u can be dropped. The restriction that u does not belong to the canonical pair containing the indices R, $R+1$ will not be significant in the case where the phase space variable label u is a spatial coordinate label, since we can then take the canonical pair labeled by the indices R, $R+1$ to lie outside the region of interest. We learn from Eq. (5.19b) that the phase space flow generated at the trace dynamics level by the generalized Poisson bracket can be represented, when averaged over the canonical ensemble and with our assumptions (1)–(4), by a unitary Heisenberg picture phase space flow of the standard quantum mechanical form, with i_{eff} playing the role of the usual i. This is the first example of how, in our framework, quantum mechanical relations are emergent from the underlying theory, without the application of *ad hoc* "canonical quantization" rules.

The assumption of vanishing ensemble trace energy can be recast as an assumption that the underlying physics is scale invariant. As we have seen in Section 2.4, in a scale-invariant theory the Lorentz trace of the trace energy-momentum tensor vanishes, $T_\mu^\mu = 0$. Since, when the τ terms can be neglected, averages over the canonical ensemble are Lorentz invariant (because $\tilde{C}$ and $\mathbf{N}$ are Lorentz invariant), we have $\langle T_{\mu\nu} \rangle_{\hat{A}V,0} = -\eta_{\mu\nu} \langle T_{00} \rangle_{\hat{A}V,0}$, and so in a scale-invariant theory $\langle \mathbf{H} \rangle_{\hat{A}V,0} = 0$. Thus, in a scale-invariant trace dynamics theory the effective Heisenberg equation of motion of Eqs. (5.19b) and (5.15c) holds when sandwiched between completely general polynomials $S_{L,R}$ constructed from the effective variables x_{eff}. Note, however, that as we also have seen in Section 2.4, scale invariance of the underlying dynamics does not imply vanishing of the Lorentz trace of an operator energy-momentum tensor, and so the emergent quantum field theory is not restricted to be scale invariant! A suggestion that this may give a mechanism for resolving the cosmological constant problem is discussed in Chapter 7. We remark that, when gravitation is included in our framework, achieving $\langle \mathbf{H} \rangle_{\hat{A}V,0} = 0$ may involve a balance between gravitational and matter contributions.

Using now the assumption that H_{eff} can be represented, in its action on the low energy (or "slow") degrees of freedom, by a suitable operator Hamiltonian function $H_{\text{eff}}(x_{r\text{eff}})$ constructed from the $x_{r\text{eff}}$, let us choose S in Eq. (5.15c) to be this function $H_{\text{eff}}(x_{r\text{eff}})$. Then substituting $\mathcal{D}$ from Eq. (5.19b) and using the chain rule we learn that

$$\langle S_L(x_{t\text{eff}}) \dot{H}_{\text{eff}} S_R(x_{t\text{eff}}) \rangle_{\hat{A}V,0}$$
$$= \langle S_L(x_{t\text{eff}}) i_{\text{eff}} \hbar^{-1} [H_{\text{eff}}, H_{\text{eff}}] S_R(x_{t\text{eff}}) \rangle_{\hat{A}V,0} = 0, \qquad (5.19c)$$

showing that within our approximations H_{eff} behaves as a constant of the motion, as required for consistency of the interpretation of Eqs. (5.19b) and (5.15c) as an effective Heisenberg dynamics. In a similar way, by use of the chain rule we find that an arbitrary polynomial function P_{eff} of the $x_{r\text{eff}}$ obeys a relation of which Eq. (5.19c) is a special case

$$\langle S_L(x_{t\text{eff}}) \dot{P}_{\text{eff}} S_R(x_{t\text{eff}}) \rangle_{\hat{A}V,0}$$
$$= \langle S_L(x_{t\text{eff}}) i_{\text{eff}} \hbar^{-1} [H_{\text{eff}}, P_{\text{eff}}] S_R(x_{t\text{eff}}) \rangle_{\hat{A}V,0}. \qquad (5.19d)$$

The effective Heisenberg picture equation of motion obtained by substituting Eq. (5.19b) into Eq. (5.15c) gives the forward time step $dt\dot{x}_{u\text{eff}}$ on the time slice on which the canonical ensemble is formed. When the underlying dynamics is time-translation invariant, a similar relation holds on each future time slice, permitting us to integrate forward with respect to time, giving the relation

$$x_{u\text{eff}}(t) = \exp(i_{\text{eff}} \hbar^{-1} H_{\text{eff}} t) x_{u\text{eff}}(0) \exp(-i_{\text{eff}} \hbar^{-1} H_{\text{eff}} t) \qquad (5.19e)$$

when sandwiched between polynomials $S_{L,R}$ and averaged over the zero-ource ensemble. (Alternatively, Eq. (5.19e) can be obtained, in Taylor expanded form, by iterating Eqs. (5.19b) and (5.15c) together with use of Eq. (5.19c), to obtain expressions for higher time derivatives of $x_{u\text{eff}}$ in terms of multiple commutators of $x_{u\text{eff}}$ with H_{eff} on the initial time slice, in analogy with the procedure used to go from Eq. (1.18c) to (1.19).) Consider now the ensemble average of an arbitrary polynomial S formed from dynamical variables taken at different times, multiplying another such polynomial S', and advance all of the variables on which S depends by the same time increment Δt, giving a time-evolved polynomial that we denote by $S|_{\Delta t}$. Then an application of Eq. (5.19e) to each variable on which S depends gives the formula

$$\langle S'S|_{\Delta t}\rangle_{\hat{\text{AV}},0} = \langle S' \exp(i_{\text{eff}}\hbar^{-1}H_{\text{eff}}\Delta t)S \exp(-i_{\text{eff}}\hbar^{-1}H_{\text{eff}}\Delta t)\rangle_{\hat{\text{AV}},0}, \qquad (5.20)$$

and similarly when S' is to the right of S.

(B) Next, let us take W in Eq. (5.19a) to be $\sigma_v x_v$, with σ_v an auxiliary c-number parameter which is a real or complex number for v bosonic, and is a real or complex Grassmann number for v fermionic, so that W in both cases is bosonic. We then have $\delta W/\delta x_s = \sigma_v \delta_{sv}$, and Eq. (5.19a) becomes (after multiplication through by i_{eff})

$$i_{\text{eff}}Dx_{u\text{eff}} = [x_{u\text{eff}}, \sigma_v x_{v\text{eff}}] - i_{\text{eff}}\hbar\omega_{uv}\sigma_v. \qquad (5.21a)$$

Equation (5.15c) then tells us that this expression vanishes when sandwiched between polynomials $S_{L,R}$ that do not contain the variable x_u, and averaged over the zero-source canonical ensemble. Moreover, the restriction on $S_{L,R}$ can be dropped if we assume that the ensemble average of the classical part of x_v vanishes, which can be achieved in the presence of sources if we assume that the source currents $j_{r\text{eff}}$ have vanishing classical parts. With these assumptions, after taking variations we have $\langle S_L(\text{Tr}x_v)S_R\rangle_{\hat{\text{AV}},0} = 0$, eliminating the unwanted source current variation terms of Eq. (5.15b). When we assume that the source currents $j_{r\text{eff}}$ have no classical parts, the arguments of the polynomials $S_{L,R}$ generated by varying these sources are correspondingly specialized to involve only the traceless parts of the effective dynamical variables $x_{r\text{eff}}$. This will be sufficient for our purposes, since the traceless parts of the dynamical variables, and not their classical parts, are the precursors of the emergent quantum degrees of freedom. The assumption of vanishing averages of the classical parts of the dynamical variables is analogous to the usual assumption of zero classical part made in field quantization in the absence of spontaneous symmetry breaking, and does not preclude dynamical spontaneous symmetry breaking, in which composites constructed from products of the fundamental dynamical variables have non-vanishing classical parts.

Translating from the compact symplectic notation used in Eq. (5.21a) (cf. also Eqs. (1.22a,b)) by recalling the definition of ω_{uv} given in Eqs. (1.16a,b), and factoring away the parameter σ_v with attention to the fact that it is Grassmann for fermionic v, we learn from Eqs. (5.15c) and (5.21a) that when multiplied by left and right polynomials $S_{L,R}$ and averaged over the zero-source ensemble, we have the effective canonical commutators

$$[q_{u\text{eff}}, q_{v\text{eff}}] = [p_{u\text{eff}}, p_{v\text{eff}}] = 0, \quad [q_{u\text{eff}}, p_{v\text{eff}}] = i_{\text{eff}}\hbar\delta_{uv} \qquad (5.21b)$$

for u, v bosonic. Similarly, we have the effective canonical anticommutators

$$\{q_{u\text{eff}}, q_{v\text{eff}}\} = \{p_{u\text{eff}}, p_{v\text{eff}}\} = 0, \quad \{q_{u\text{eff}}, p_{v\text{eff}}\} = i_{\text{eff}}\hbar\delta_{uv} \qquad (5.21c)$$

for u, v fermionic, with the corresponding commutators of bosonic with fermionic quantities all vanishing. In other words, the entire canonical algebra for boson and fermion degrees of freedom is compactly encoded in the statement that the right-hand side of Eq. (5.21a) vanishes. Thus, as suggested at the beginning of this chapter, the Ward identities do in fact lead to an equipartitioning of $\tilde{C}$, giving rise in an emergent fashion to the canonical commutator/anticommutator structure that is the basis for field quantization.

Note that the emergence of the canonical algebra implicitly requires a limiting process. As is well known, in a complex Hilbert space the canonical algebra $[q, p] = i\hbar$, $[q, i] = [p, i] = 0$ cannot have finite-dimensional (or more generally, trace class) representations, since when cyclic permutation under the trace is allowed, the trace of $[q, p]$ is zero while the trace of i is iN, with N the dimension of the Hilbert space. Since our emergent Heisenberg algebra in the bosonic case has the form $[q_{\text{eff}}, p_{\text{eff}}] = i_{\text{eff}}\hbar$, and since $\text{Tr} i_{\text{eff}} = 0$, a contradiction does not arise at the level of a single commutator. However, let us consider a relation derived by using the canonical commutator $[q, p] = i\hbar$ twice, for example, the relation $q^2 p^2 + p^2 q^2 - 2qp^2 q = -2\hbar^2$. Now there is a conflict with the trace class assumption, even when the role of i is played by i_{eff}, since assuming cyclic permutation under the trace, the left-hand side of this identity would have a vanishing trace, while the right-hand side has non-vanishing trace $-2N\hbar^2$. However, it is consistent for the canonical algebra to emerge as the limit $N \to \infty$ of a matrix algebra in an N-dimensional Hilbert space, or as an idealized approximation to a matrix algebra in a Hilbert space with N large but finite, or as an approximation to a more general type of trace class algebra.

(C) Finally, let us take W in Eq. (5.19a) to be a general self-adjoint polynomial G, so that **G** is the generator of general canonical transformations of the trace dynamics as described in Eqs. (2.13a,b). Then combining Eq. (5.19a) with Eq. (2.13b), and dividing by $\hbar$, we get

$$\hbar^{-1}\mathcal{D}x_{u\text{eff}} = i_{\text{eff}}\hbar^{-1}[G_{\text{eff}}, x_{u\text{eff}}] - \delta x_{u\text{eff}}. \tag{5.22a}$$

This tells us that sandwiched between polynomials $S_{L,R}$ that do not contain the variable x_u, and averaged over the zero-source canonical ensemble, general infinitesimal canonical transformations are effectively generated by unitary transformations of the form

$$U_{\text{can eff}} = \exp(i_{\text{eff}}\hbar^{-1}G_{\text{eff}}). \tag{5.22b}$$

Moreover, when $\mathbf{W}$ is a conserved extensive generator $\mathbf{G}$ that is a symmetry of the canonical ensemble, so that $\mathbf{G}$ effectively vanishes inside canonical averages, then the restriction on $S_{L,R}$ can be dropped and Eq. (5.22b) holds as a general "off-shell" relation. Examples of conserved generators satisfying this condition are the trace momentum and trace angular momentum, which have vanishing generalized Poisson bracket with the trace Hamiltonian (as well as with the Poincaré invariants $\mathbf{N}$ and $\tilde{C}$; see remarks below). The trace boost generators are linearly increasing with time, because for a Poincaré invariant underlying theory, their generalized Poisson brackets with $\mathbf{H}$ are proportional to the trace momentum generators, which are conserved. Since we have assumed a canonical ensemble that is at rest and thus defines no preferred spatial vector, the canonical ensemble averages of the trace boost generators are therefore also zero, and so the boosts will also be represented by unitary (or when spinors are present, pseudounitary) transformations. In trace dynamics models defined as continuum theories, such as the examples of Sections 2.3 and 2.4 and Chapter 3, we therefore see that the canonical trace Poincaré generators are represented at the effective level by operator generators. In particular, in Poincaré invariant theories the trace three-momentum will have a corresponding operator three-momentum $\vec{P}_{\text{eff}}$ that is time independent (by similar arguments to those used to show that H_{eff} is time independent), and that generates translations involving spatial displacement $\Delta\vec{x}$ through a unitary transformation

$$U_{\text{trans eff}}(\Delta\vec{x}) = \exp(i_{\text{eff}}\hbar^{-1}\Delta\vec{x} \cdot \vec{P}_{\text{eff}}). \tag{5.22c}$$

In a similar manner, internal symmetries with generators $\mathbf{G}$ that are symmetries of the trace Hamiltonian will have $\mathbf{G}$ effectively vanishing inside canonical averages, and so will also be represented by unitary transformations in the effective quantum theory. Finally, let $\mathbf{G} = \mathbf{G}_\Lambda = \text{Tr}\Lambda\tilde{C}$ be the conserved generator constructed in Eq. (2.11a) as the generator of global unitary transformations, with the restriction now that Λ should be a matrix that commutes with i_{eff}, and should have vanishing trace in each of the two K-dimensional subspaces obtained by reducing the Hilbert space modulo i_{eff} as in Section 4.5. This generator $\mathbf{G}$ vanishes inside canonical averages, since these restrictions enforce the condition $\text{Tr}\, i_{\text{eff}}\Lambda = 0$.

Thus Eq. (5.22b) holds as an off-shell relation in this case also, and corresponds to the global unitary transformations U_{eff} that produce nontrivial unitary changes of basis in the emergent quantum theory, since the restrictions that we have imposed exclude only those trivial transformations that act as an overall phase times the unit matrix in each the K-dimensional subspaces on which $i_{\text{eff}} = \pm i$.

To sum up, we have shown that the basic structure of quantum field theory – the canonical commutator/anticommutator structure, time evolution in the Heisenberg picture, and the unitary generation of canonical transformations – emerges from the statistical thermodynamics of matrix models with a global unitary invariance. In the above discussion the polynomials $S_{L,R}$ were constructed from phase space variables (or at least, their traceless or non-classical parts) taken at the common time used to form the canonical ensemble. However, once we have ascertained that there are no restrictions on the labels of these variables (other than that they cannot contain x_R, x_{R+1}, which can be taken to lie outside all relevant causal horizons), pieces of the polynomials $S_{L,R}$ can be used to construct powers of the effective Hamiltonian H_{eff}, and the formal integration of the Heisenberg picture dynamics given in Eqs. (5.19e) and (5.20) can be used to extend our results to identities sandwiched between general polynomial functions of the phase space variables taken at general times. This extends the identities "off-shell" by giving them the status of "weak" operator relations, that is relations that hold when averaged over all smooth test functions formed from the traceless parts of the dynamical variables.

The results that we have obtained suggest a correspondence between operator polynomials in the underlying trace dynamics, and operator polynomials in complex quantum field theory

$$S(\{x_{r\text{eff}}\}) \Leftrightarrow S(\{X_{r\text{eff}}\}), \tag{5.23a}$$

with the $X_{r\text{eff}}$ on the right quantized operators in a quantum theory with the role of i played by the matrix i_{eff}, which commutes with all of the $X_{r\text{eff}}$. Since i_{eff} diagonalizes into two $K \times K$ blocks as indicated in Eq. (4.40), the correspondence of Eq. (5.23a) actually gives two uncoupled copies of a complex quantum field dynamics, on one of which i_{eff} acts as i and on the other of which i_{eff} acts as the complex conjugate $-i$. Let us now focus on the copy on which i_{eff} acts as i; similar considerations apply to the copy on which i_{eff} acts as $-i$. (Some speculations on possible physical implications of the existence of two copies of the emergent quantum theory, particularly with regard to cosmology, are given in Chapter 7.)

We now make two further structural assumptions:

(5) In the continuum limit, with the indices r labeling infinitesimal spatial boxes as well as internal symmetry structure, the underlying trace Lagrangian **L** is Poincaré invariant.

(6)   The underlying dynamics leads to an H_{eff} that is bounded from below by the magnitude of the corresponding effective three-momentum operator $\vec{P}_{\text{eff}}$, and there is a unique eigenvector ψ_0 with lowest eigenvalue of H_{eff} and zero eigenvalue of $\vec{P}_{\text{eff}}$. (Although introduced here as an independent assumption, the existence of a lowest eigenvector may follow from the boundedness assumption on H_{eff}, and this in turn may be deducible from an analogous boundedness assumption at the trace dynamics level, stating that $\mathbf{H}$ is bounded from below by $(\vec{\mathbf{P}} \cdot \vec{\mathbf{P}})^{\frac{1}{2}}$.)

With these two assumptions we can make contact with quantum field theory. According to assumption (6), the eigenvector ψ_0 acts as the conventional vacuum state, and we have seen that within ensemble averages, the underlying matrices obey properties analogous to those of quantum fields. With this motivation, we propose the following precise correspondence between trace dynamics canonical ensemble averages and Wightman functions in the emergent quantum field theory

$$\psi_0^\dagger \langle S(\{x_{r\text{eff}}\})\rangle_{\widehat{\text{AV}}} \psi_0 = \langle \text{vac}|S(\{X_{r\text{eff}}\})|\text{vac}\rangle. \qquad (5.23\text{b})$$

(Equation (5.23b) applies to xs that are bosonic and fermionic canonical qs and ps. Thus in the fermionic case, $q_{r\text{eff}} \leftrightarrow Q_{r\text{eff}}$ and $p_{r\text{eff}} \leftrightarrow P_{r\text{eff}}$. However, when the fermionic quantized operators are rewritten in terms of $\Psi_{r\text{eff}}$ and $\Psi_{r\text{eff}}^\dagger$, the correspondence reads $\psi_{r\text{eff}} \leftrightarrow \Psi_{r\text{eff}}$ and $\psi_{r\text{eff}}^\dagger \leftrightarrow i_{\text{eff}}\Psi_{r\text{eff}}^\dagger$, since the correct definition of the operator $\Psi_{r\text{eff}}^\dagger$ is $P_{r\text{eff}} = i_{\text{eff}}\Psi_{r\text{eff}}^\dagger$, as discussed in further detail in Section 5.5.) When the underlying trace dynamics is Poincaré invariant, the fact that Poincaré transformations are canonical transformations with global unitary invariant generators implies, by the discussion of Section 2.2, that $\tilde{C}$ is a Poincaré invariant. Hence the term $\text{Tr}\tilde{\lambda}\tilde{C}$ in the canonical ensemble is a Poincaré invariant, and in particular is a Lorentz scalar. By a similar argument, the trace fermion number $\mathbf{N}$ is a Poincaré invariant. Although the term $\tau\mathbf{H}$ in the canonical ensemble is *not* a Lorentz scalar, in the low energy regime where approximation (1) is valid, so that variations in the τ term can be neglected in the Ward identities, the Lorentz noninvariance associated with this term decouples, giving a Lorentz invariant effective field theory. Hence with the assumptions and approximations that we have made, the correspondence of Eq. (5.23b) defines Wightman functions of a Lorentz invariant effective field theory, as we shall now outline.

Let us briefly enumerate the basic properties of Wightman functions (Streater and Wightman, 1968) that are needed to reconstruct local quantum field theory (see Appendix F for details) and indicate why, with the correspondence of Eq. (5.23b), it is plausible that they are obeyed. We assume for this argument that the subscript r on the phase space variables x_r labels infinitesimal spatial boxes, so that we are discussing the trace dynamics analog of a continuum field theory.

(i) By constructing the Wightman functions from thermodynamic averages, they are expected to have the requisite smoothness properties, i.e., they should be tempered distributions.

(ii) Lorentz covariance of the Wightman functions follows from Lorentz invariance of $\tilde{C}$ and $\mathbf{N}$ and our assumption that the noninvariant $\tau\mathbf{H}$ terms in the canonical ensemble effectively decouple from the low energy theory.

(iii) The spectral condition for the Wightman functions follows from the formulas of Eqs. (5.19e), (5.20), (5.22c), and the analog of Eq. (5.20) for spatial translations (which combined give the effect, within canonical ensemble averages, of space-time translation of the phase space variables), together with the spectral assumptions stated in (6).

(iv) Local commutativity of the Wightman functions follows from the canonical commutation/anticommutation relations derived, in the approximation of neglecting the τ terms, in Eqs. (5.21a–c), together with Lorentz covariance, which allows us to transform any spacelike separated pair of variables into spatially separated variables on the same time slice.

(v) The hermiticity and positivity properties of the Wightman functions follow directly from the correspondence with thermodynamics averages of Eq. (5.23b), since these averages have analogous properties.

(vi) The cluster property of the Wightman functions corresponds to the assumptions about $\mathbf{H}$ made in the clustering argument used in the discussion of the microcanonical ensemble in Eqs. (4.19a–c).

Once the properties (i)–(vi) are obeyed, the fundamental reconstruction theorem of Wightman shows that a local, relativistic quantum field theory can be reconstructed from the Wightman functions. Thus, under the approximations leading to the correspondence of Eq. (5.23b), the canonical ensemble averages of trace dynamics give rise, as an effective low energy dynamics, to a local relativistic quantum field theory.

In Section 4.5, we pointed out that when statistical averages are calculated with a unitary fixing that eliminates the residual global unitary invariance group commuting with i_{eff}, then only averages of trace functions of the matrix variables are independent of the choice of fixing. This means that quantities like $\langle S(\{x_{r\text{eff}}\})\rangle_{\text{AV}}$ depend on the choice of unitary fixing. However, let us write the ground state projector $\Pi_0 = \psi_0\psi_0^\dagger$ as a Cauchy integral over the resolvent for H_{eff}

$$\Pi_0 = \frac{1}{2\pi i}\int dz \frac{1}{z - H_{\text{eff}}}, \qquad (5.23c)$$

with the contour an infinitesimal counterclockwise circle around $z = 0$. Then the left-hand side of Eq. (5.23b) can be rewritten as

$$\langle \text{Tr} \frac{1}{2\pi i}\int dz \frac{1}{z - H_{\text{eff}}} S(\{x_{r\text{eff}}\})\rangle_{\text{AV}}, \qquad (5.23d)$$

which has the form of the average of a trace polynomial. Therefore the emergent Wightman functions are independent of the choice of unitary fixing. More generally, the measurable quantities in quantum theory are transition probabilities, which can be written as (see the discussion of the Jordan formulation of quantum theory in Gürsey and Tze, 1996)

$$|\langle \beta | \Omega | \alpha \rangle|^2 = \text{Tr} \Omega \Pi_\beta \Omega \Pi_\alpha, \qquad (5.23e)$$

with Ω some self-adjoint operator. Since Π_β and Π_α can be written as contour integrals over the resolvents of the corresponding operators for which they are spectral projectors, the transition probabilities in quantum theory can always be written as trace functionals of operators, and so are also physical observables within our framework.

To conclude this section, we note that there is a natural hierarchy of matrix structures leading from the underlying trace dynamics, to the emergent effective complex quantum field theory, to the classical limit, as drawn in Fig. 2 of the Introduction. In the underlying theory, the matrices x_r are of completely general structure. No commutation properties of the x_r are assumed at the trace dynamics level, and since all degrees of freedom communicate with one another, the dynamics is completely nonlocal. In the effective quantum theory, the x_r are still matrices, but with a restricted structure that obeys the canonical commutator/anticommutator algebra. Thus, locality is an emergent property of the effective theory, even though it is not a property of the underlying trace dynamics. Finally, in the limit in which the matrices x_r are dominated by their c-number or classical parts defined in Eqs. (2.16a,b), the effective quantum field dynamics becomes an effective classical dynamics. Thus, both classical mechanics and quantum mechanics are subsumed in the more general trace dynamics, as reflected in the following hierarchy of matrix structures, corresponding to increasing specialization: general $\rightarrow$ canonical quantum plus c-number $\rightarrow$ c-number, classical. The Heisenberg uncertainty relations characteristic of the canonical algebra of quantum mechanics appear only at the middle level of this hierarchy of matrix structures. At the bottom or trace dynamics level, there is no indeterminacy in the dynamics of the operator matrix elements (at least in principle, given the initial conditions), because no dynamical information has been discarded. Although one can form analogs of the Heisenberg uncertainty relations for expectations of operator variances at the trace dynamics level, they will involve unknown and effectively random operator commutators in place of the usual quantum mechanical $\frac{1}{2}\hbar$. At the top or classical level, there are no apparent indeterminacies because quantum uncertainties are masked by large system size.

5.4 Restrictions on the underlying theory implied by further Ward identities

In this section we shall take a critical look at the assumptions and approximations made in the previous section in the course of our argument for the emergence of quantum behavior. We shall do this by deriving further Ward identities, and showing that their consistency with the Ward identity derived in Section 5.1, and with the approximations made in Section 5.3, places nontrivial constraints on the structure of the underlying trace dynamics. In particular, these constraints take the form of special support properties in operator phase space, and a requirement of balance between the numbers of bosonic and fermionic degrees of freedom.

The first additional Ward identity is obtained by proceeding as we did starting from Eq. (5.5a), but with the factor $\{\tilde{C}, i_{\text{eff}}\} = 2i_{\text{eff}}\tilde{C}_{\text{eff}}$ appearing in the trace $\text{Tr}\{\tilde{C}, i_{\text{eff}}\}W$ replaced at the outset by a constant, which we can take to be unity. Thus, we start now from

$$Z_j \langle \text{Tr}W \rangle_{\widehat{\text{AV}},j} = \int d\hat{\mu} \exp\left(-\text{Tr}\tilde{\lambda}\tilde{C} - \tau\mathbf{H} - \eta\mathbf{N} - \sum_r \text{Tr}j_r x_r\right)\text{Tr}W, \quad (5.24a)$$

and proceed as we did in Section 5.1. The derivation completely parallels that leading to Eqs. (5.14a,b), except that the term corresponding to Eq. (5.9b), which comes from the variation of the factor $\tilde{C}$ in $\text{Tr}\{\tilde{C}, i_{\text{eff}}\}W$, is absent. Dropping this term from Eq. (5.14b), replacing $\{\tilde{C}, i_{\text{eff}}\}$ on the right-hand side by $-2\hbar$ and factoring away $-\hbar$, and using Eq. (5.10b) to simplify the final term on the right, we get

$$\langle \Lambda'_{u\text{eff}} \rangle_{\widehat{\text{AV}},j} = 0, \quad (5.24b)$$

with $\Lambda'_{u\text{eff}}$ given by

$$\Lambda'_{u\text{eff}} = \left(-\tau\dot{x}_{u\text{eff}} + i\eta\xi_u x_{u\text{eff}} - \sum_s \omega_{us}j_{s\text{eff}}\right)\mathbf{W}$$

$$+ \sum_s \omega_{us}\left(\frac{\delta\mathbf{W}}{\delta x_s}\right)_{\text{eff}}. \quad (5.24c)$$

Correspondingly, after variation of source terms, instead of Eq. (5.16b) we now get (for monomial S, S_L, S_R)

$$\langle S_L(x_{t\text{eff}})\sum_u S(x_{r\text{eff}}, r \neq u; \mathcal{D}'x_{u\text{eff}})S_R(x_{t\text{eff}})\rangle_{\widehat{\text{AV}},0} = 0, \quad (5.25a)$$

with $\mathcal{D}'x_{u\text{eff}}$ given by

$$\mathcal{D}'x_{u\text{eff}} = (-\tau \dot{x}_{u\text{eff}} + i\eta\xi_u x_{u\text{eff}})\mathbf{W}$$

$$+ \sum_s \omega_{us} \left(\frac{\delta\mathbf{W}}{\delta x_s}\right)_{\text{eff}}. \tag{5.25b}$$

It is apparent from Eq. (5.25b) that we *cannot* make an approximation of neglecting both the η term and the τ term on the right-hand side, since this would lead to the incorrect conclusion, for example, that

$$\left\langle \left(\frac{\delta\mathbf{W}}{\delta x_s}\right)_{\text{eff}} \right\rangle_{\hat{A}\hat{V},0} = 0 \tag{5.26a}$$

for general W. In other words, assumption (1) of Section 5.3, which we recall states that

$$\tau \dot{x}_{u\text{eff}} \text{Tr}\tilde{C}_{\text{eff}} i_{\text{eff}} W_{\text{eff}} = \tau \dot{x}_{u\text{eff}} \text{Tr}\tilde{C}_{\text{eff}} i_{\text{eff}} W \tag{5.26b}$$

can be neglected, cannot be extended to the assumption that the corresponding expression obtained by replacing $\tilde{C}_{\text{eff}}$ by its ensemble average $i_{\text{eff}}\hbar$ can also be neglected. This does not invalidate the reasoning of Section 5.3, but does place constraints on the support structure of the underlying trace dynamics. To give a simple illustration, if f and g are non-negative functions, the vanishing of the average $(fg)_{\text{AV}}$ over a domain including the supports of f and g does not contradict the fact that $((f)_{\text{AV}}g)_{\text{AV}} = (f)_{\text{AV}}(g)_{\text{AV}}$ is non-zero and positive; what is necessary and sufficient to achieve the vanishing of the former is for f and g to have nonintersecting domains of support, so that $f = 0$ where $g > 0$ and vice versa. We are faced with a similar situation with respect to $\tilde{C}_{\text{eff}}$, which although not of definite sign has, by assumption, a non-vanishing canonical ensemble average $i_{\text{eff}}\hbar$. A sufficient condition for us to be able to neglect Eq. (5.26b), without contradicting Eq. (5.25b), is for $\dot{x}_{u\text{eff}}$ and $\tilde{C}_{\text{eff}}$ to have disjoint domains of support, which is why we have phrased assumption (1) in terms of support properties.

At the same time, we cannot impose this property by requiring the stronger condition that $x_{u\text{eff}}$ and $\tilde{C}_{\text{eff}}$ have disjoint support, since this would contradict assumption (3), which states that $\{\tilde{C}, i_{\text{eff}}\}$ can be replaced by its ensemble average in the final term in Eq. (5.15a). Thus, what is required is that the "slow" components of $x_{u\text{eff}}$, for which $\tau \dot{x}_{u\text{eff}}$ is effectively zero, share a common domain of support with $\tilde{C}_{\text{eff}}$, and be slowly varying with respect to the scale of variations of $\tilde{C}_{\text{eff}}$, so that in "slow" terms in the Ward identity $\tilde{C}_{\text{eff}}$ is effectively equal to its ensemble average. At the same time, we must require that the "fast" components of $x_{u\text{eff}}$, for which $\dot{x}_{u\text{eff}}$ is significant, have disjoint support from $\tilde{C}_{\text{eff}}$, so that the ensemble average of their product is effectively zero. To see that these requirements are compatible, and suffice to do what is needed, let us write

$$x_{r\text{eff}} = x_{r\text{eff}}^{\text{slow}} + x_{r\text{eff}}^{\text{fast}}, \tag{5.27a}$$

and postulate that $\tilde{C}_{\text{eff}}$ has disjoint support from $x_{r\text{eff}}^{\text{fast}}$. Then in the final term of Eq. (5.15a) we have

$$\left(W_s^{R\ell} \frac{1}{2}\{\tilde{C}, i_{\text{eff}}\} W_s^{L\ell} \right)_{\text{eff}} = \left(W_s^{R\ell\,\text{slow}} \frac{1}{2}\{\tilde{C}, i_{\text{eff}}\} W_s^{L\ell\,\text{slow}} \right)_{\text{eff}}, \tag{5.27b}$$

so that the "fast" terms do not appear in this expression, and we can then apply assumption (3) to replace $\tilde{C}_{\text{eff}}$ by $i_{\text{eff}}\hbar$ in this term. Similarly, by the assumed support properties

$$\tau \dot{x}_{r\text{eff}} \text{Tr} \tilde{C}_{\text{eff}} i_{\text{eff}} W_{\text{eff}} = \tau \dot{x}_{r\text{eff}}^{\text{slow}} \text{Tr} \tilde{C}_{\text{eff}} i_{\text{eff}} W_{\text{eff}}^{\text{slow}} \simeq 0, \tag{5.27c}$$

but on the other hand

$$\tau \dot{x}_{r\text{eff}} \text{Tr} W = \tau \dot{x}_{r\text{eff}} \text{Tr} W_{\text{eff}} = \tau (\dot{x}_{r\text{eff}}^{\text{slow}} + \dot{x}_{r\text{eff}}^{\text{fast}}) \text{Tr} (W_{\text{eff}}^{\text{slow}} + W_{\text{eff}}^{\text{fast}})$$
$$\simeq \tau \dot{x}_{r\text{eff}}^{\text{fast}} \text{Tr} (W_{\text{eff}}^{\text{slow}} + W_{\text{eff}}^{\text{fast}}) \neq 0, \tag{5.27d}$$

so the additional Ward identity derived in Eq. (5.25b) (with the η term neglected) can be satisfied.

More specific statements of our assumptions (1) and (3) can be given if we specialize the underlying trace dynamics (i) to the supersymmetric case, where the numbers n_B and n_F of bosonic and fermionic degrees of freedom are equal, and (ii) to the specific case of supersymmetric Yang–Mills theory, where we have seen in Section 3.2 that $\tilde{C}$ reduces to a surface integral at spatial infinity. As preparation for these specializations, let us follow the notation of Eq. (4.41b) and write $x_r = x_{r\text{eff}} + x_{r\,12}$, which when substituted into Eq. (5.1) for $\tilde{C}$ gives

$$\tilde{C} = \sum_{r,s} \omega_{rs} (x_{r\text{eff}} x_{s\text{eff}} + x_{r\text{eff}} x_{s\,12} + x_{r\,12} x_{s\text{eff}} + x_{r\,12} x_{s\,12}). \tag{5.28a}$$

Taking the effective projection, and using Eq. (4.41c), which implies that $(x_{s\,12})_{\text{eff}} = (x_{r\,12})_{\text{eff}} = 0$ and that $(x_{r\,12} x_{s\,12})_{\text{eff}} = x_{r\,12} x_{s\,12}$ (since i_{eff} anticommutes with $x_{r\,12}$ and with $x_{s\,12}$, it commutes with their product), we get

$$\tilde{C}_{\text{eff}} = \sum_{r,s} \omega_{rs} (x_{r\text{eff}} x_{s\text{eff}} + x_{r\,12} x_{s\,12}). \tag{5.28b}$$

The first term on the right-hand side in Eq. (5.28b) can be rewritten as

$$\sum_{r,s} \omega_{rs} x_{r\text{eff}} x_{s\text{eff}} = \sum_{r \in B} [q_{r\text{eff}}, p_{r\text{eff}}] - \sum_{r \in F} \{q_{r\text{eff}}, p_{r\text{eff}}\}. \tag{5.28c}$$

Within ensemble averages, we have seen that the commutators and anticommutators in Eq. (5.28c) have the effective canonical values given in Eqs. (5.21b,c), and so within averages Eq. (5.28c) is effectively $(n_B - n_F)i_{\text{eff}}\hbar$, and we have

$$\tilde{C}_{\text{eff}} \simeq (n_B - n_F)i_{\text{eff}}\hbar + \sum_{r,s} \omega_{rs} x_{r\,12} x_{s\,12}, \qquad (5.28d)$$

with n_B and n_F respectively the numbers of bosonic and fermionic matrix degrees of freedom.

We can now apply this equation in two ways. First of all, within an ensemble average containing no other factors, Eq. (5.28d) becomes

$$i_{\text{eff}}\hbar \simeq (n_B - n_F)i_{\text{eff}}\hbar + \left\langle \sum_{r,s} \omega_{rs} x_{r\,12} x_{s\,12} \right\rangle_{\hat{\text{AV}},0}. \qquad (5.29a)$$

This shows that if the difference $n_B - n_F$ becomes infinite, then there is an inconsistency unless the second term in Eq. (5.29a) becomes large in such a way as to cancel the infinite part of the first term. This is implausible, and so we conclude that consistency of our approximations requires that $n_B - n_F$ should be finite, even when n_B becomes large. (A more precise version of this argument for boson–fermion balance will be given shortly.) Specializing now to the case of supersymmetric theories, for which n_B is exactly equal to n_F, the first term on the right in Eq. (5.28d) vanishes, and we get

$$\tilde{C}_{\text{eff}} \simeq \sum_{r,s} \omega_{rs} x_{r\,12} x_{s\,12}. \qquad (5.29b)$$

From this equation, we see that the support properties stated in the lines preceding Eq. (5.27a), that were expressed in terms of $\tilde{C}_{\text{eff}}$, can now be expressed in terms of support properties involving $x_{s\,12}$. Thus, a sufficient condition for the needed support properties is that $x_{u\text{eff}}^{\text{fast}}$ should have disjoint support from $x_{s\,12}$ for all u, s, while $x_{u\text{eff}}^{\text{slow}}$ should have a common support with $x_{s\,12}$, in such a way that $\tilde{C}_{\text{eff}}$ can be replaced by its ensemble average in expressions involving "slow" quantities. From Eq. (5.29a) with $n_B - n_F = 0$, we see that $i_{\text{eff}}\hbar$ obeys the sum rule

$$i_{\text{eff}}\hbar \simeq \left\langle \sum_{r,s} \omega_{rs} x_{r\,12} x_{s\,12} \right\rangle_{\hat{\text{AV}},0}. \qquad (5.29c)$$

We have learned from this analysis that the quantities for which disjoint support is required are distinct components under the separation into "eff" and "12" components introduced in Eq. (4.41b). A further distinction is present in supersymmetric Yang–Mills theories, in which we have seen that $\tilde{C}$ reduces to a surface term at spatial infinity, giving the emergence of quantum mechanics in these theories a "holographic" flavor. (For a review of recent ideas on a possible holographic structure of physical theories, see Bousso, 2002.) The degrees of freedom $x_{r\,(\text{vol})}$ which describe observable physics are then degrees of freedom residing in the interior volume, whereas the components $x_{r\,(\text{surf})\,12}$ entering into Eq. (5.29b) for $\tilde{C}$ reside on the surface at infinity. For such theories, the required support properties state that (i) for volume degrees of freedom, $x^{\text{fast}}_{r(\text{vol})\text{eff}}$ should have disjoint phase space support from degrees of freedom $x_{r(\text{surf})\,12}$ residing on the surface at spatial infinity, so that $\tau \dot{x}_{r(\text{vol})\text{eff}} \text{Tr} \tilde{C}_{\text{eff}} i_{\text{eff}} W_{\text{eff}}$ can be neglected even when $\tau \dot{x}_{r(\text{vol})\text{eff}} \text{Tr} W_{\text{eff}}$ is significant; and (ii) the volume degrees of freedom $x^{\text{slow}}_{r(\text{vol})\text{eff}}$ should have a common support with the degrees of freedom $x_{r(\text{surf})\,12}$ residing on the surface at infinity and should be slowly varying relative to $\tilde{C}_{\text{eff}}$, so that we can freely replace $\tilde{C}_{\text{eff}}$ by its ensemble average $i_{\text{eff}}\hbar$ inside averages of products of factors $x^{\text{slow}}_{r(\text{vol})\text{eff}}$. (Recall that the distinction between volume and surface here refers only to operator labels r, and does not directly translate into support properties in the operator phase space.) These statements are as far as we have been able to carry a general analysis of the needed support properties. A further understanding of whether they can be realized will require a study of specific models for the underlying trace dynamics; constructing realistic candidates requires a solution to the problem of obtaining a large hierarchy of scales between the underlying "fast" and the phenomenological "slow" degrees of freedom, which was invoked as part of the justification for neglecting the τ terms in the Ward identities. This problem is the same as the so-called hierarchy problem of particle unification models, and its solution may require deep new physical ideas. At a minimum, what our analysis has accomplished is to show that the support properties needed for assumptions (1) and (3) are not contradictory, and so cannot be used in any obvious way to construct an argument contradicting our program.

The support properties required for the emergence of quantum behavior can also be characterized in physical terms as the requirement that the canonical ensemble should possess a certain "rigidity", in the sense that the contribution of the ensemble variation $(\delta \rho / \delta x_s)_{\text{eff}}$ to the Ward identity of Section 5.1 can be neglected, as posited in assumptions (1) and (2) of Section 5.3. The need for a rigid statistical ensemble in our context suggests a possible analogy with the concept of London rigidity in the theory of superconductivity. In the presence of an applied vector potential $\vec{A}$, the induced current density $\vec{j}$ in a metal is given by

$$\langle \vec{j} \rangle = -\frac{ne}{m} \langle \vec{p} + e\vec{A} \rangle, \tag{5.29d}$$

with $n, m, e, \vec{p}$ respectively the electron density, mass, charge, and three-momentum operator. In a normal metal the two terms on the right-hand side of Eq. (5.29d) nearly cancel, leaving a small residual diamagnetism. However, in a superconductor the rigidity of the wave function leads to the vanishing of $\langle \vec{p} \rangle$, giving perfect diamagnetism and the Meissner effect. An analogy with the analysis of this section would equate normal metal behavior with the case in which $\tilde{C}_{\text{eff}}$ can be replaced by its ensemble average in all terms, including the τ term, in the

Ward identities. In this case $-\hbar$ times the right-hand side of Eq. (5.24c) is equal to the non-commutator part of the right-hand side of Eq. (5.15a) when the sources j_s are set equal to zero, leading to vanishing of the emergent canonical commutator/anticommutator and to an effective classical dynamics. Similarly, the analogy would equate superconducting behavior with the case in which the τ term containing $\tilde{C}_{\text{eff}}$ can be dropped because of "rigidity" of $(\delta\rho/\delta x_s)_{\text{eff}}$, leading as seen in Section 5.3 to an emergent canonical commutator/anticommutator as an analog of the superconductive Meissner effect. In this analogy, the Planck energy and the associated energy scale hierarchy would play the role of the superconductive energy gap. We suggest that this analogy may be useful in identifying the particular underlying trace dynamics for which the assumptions needed for emergent quantum mechanics are realized.

We turn now to deriving a more precise statement (Adler and Kempf, 1998) of the requirement of boson–fermion balance, which we achieve by deriving yet another Ward identity. This is obtained by proceeding as we did starting from Eq. (5.5a), but with $\{\tilde{C}, i_{\text{eff}}\}$ replaced by $\tilde{C}$ in the trace factor $\text{Tr}\{\tilde{C}, i_{\text{eff}}\}W$, so that this factor is taken now as $\text{Tr}\tilde{C}W$; since $\text{Tr}\{\tilde{C}, i_{\text{eff}}\}W = \text{Tr}\tilde{C}\{i_{\text{eff}}, W\}$, this is equivalent to replacing $2i_{\text{eff}}W_{\text{eff}} = \{i_{\text{eff}}, W\}$ by W. Making these replacements in Eq. (5.11b), we get the Ward identity

$$\langle \Sigma_s'' \rangle_{\hat{\text{AV}}, j} = 0, \tag{5.30a}$$

with Σ_s'' given by

$$\Sigma_s'' = \left(-\left[\tilde{\lambda}, \sum_r \omega_{rs} x_r\right] - \tau \sum_r \omega_{rs} \dot{x}_r - i\eta \sum_r \tilde{\omega}_{rs} x_r - j_s\right) \text{Tr}\tilde{C}W \tag{5.30b}$$

$$+ \left[W, \sum_r \omega_{rs} x_r\right] + \sum_\ell \epsilon_\ell W_s^{R\ell} \tilde{C} W_s^{L\ell}.$$

We now follow a different procedure from that used in Section 5.1, by immediately taking the sources j to vanish, and by not taking an overall effective projection. The contribution to the Ward identity of the first term on the right-hand side of Eq. (5.30b) can be rewritten as

$$-\left[\tilde{\lambda}, \langle \sum_r \omega_{rs} x_r \text{Tr}\tilde{C}W \rangle_{\hat{\text{AV}}, 0}\right], \tag{5.30c}$$

which vanishes by virtue of Eq. (4.47a). Multiplying by ω_{us} and summing over s, and evaluating the sums using Eqs. (5.13a,b), in place of Eqs. (5.14a,b) we now get the Ward identity

$$\langle \Lambda_u'' \rangle_{\hat{\text{AV}}, 0} = 0, \tag{5.31a}$$

with Λ_u'' given by

$$
\Lambda_u'' = (-\tau \dot{x}_u + i\eta \xi_u x_u)\text{Tr}\tilde{C}W
$$
$$
+ [W, x_u] + \sum_{s,\ell} \omega_{us}\epsilon_\ell W_s^{R\ell}\tilde{C}W_s^{L\ell}. \tag{5.31b}
$$

Let us now apply this to the particular choice $W = \sigma_t x_t$, with σ_t again an auxiliary c-number parameter which is a complex number for t bosonic, and a complex Grassmann number for t fermionic. Then (cf. Eqs. (5.10b,c)) we have only one term in the sum over ℓ, with $\epsilon_{\ell=1} = 1$, $W_s^{L\ell=1} = \sigma_t \delta_{st}$, and $W_s^{R\ell=1} = 1$, so that the final term in Eq. (5.31b) reduces to

$$
\omega_{ut}\tilde{C}\sigma_t, \tag{5.31c}
$$

and has the ensemble average (continuing to assume that $\langle \tilde{C} \rangle_{\hat{\text{AV}}} = \langle \tilde{C} \rangle_{\text{AV}}$)

$$
\omega_{ut} i_{\text{eff}} \hbar \sigma_t. \tag{5.31d}
$$

Hence in this special case the Ward identity of Eqs. (5.31a,b) reduces to

$$
0 = \langle (-\tau \dot{x}_u + i\eta \xi_u x_u)\text{Tr}\tilde{C}\sigma_t x_t \rangle_{\hat{\text{AV}},0} + \langle [\sigma_t x_t, x_u] \rangle_{\hat{\text{AV}},0} + \omega_{ut} i_{\text{eff}} \hbar \sigma_t. \tag{5.32a}
$$

Since the final two terms in Eq (5.32a) are manifestly traceless, and since $\tilde{C}$ is traceless, projecting out the traceless part of the first term using the notation of Eq. (2.16a), and rearranging terms, we arrive at (Adler and Kempf, 1998)

$$
\langle [x_u, \sigma_t x_t] \rangle_{\hat{\text{AV}},0} = i_{\text{eff}} \hbar \omega_{ut} \sigma_t + \langle (-\tau \dot{x}_u' + i\eta \xi_u x_u')\text{Tr}\tilde{C}\sigma_t x_t' \rangle_{\hat{\text{AV}},0}. \tag{5.32b}
$$

Letting the indices t and u in Eq. (5.32b) be either both bosonic or both fermionic, and in the fermionic case, for simplicity, setting the "chemical potential" η equal to zero (it is easy to extend the analysis to $\eta \neq 0$), we get by referring to Eqs. (1.16a,b) the respective relations

$$
\langle [q_r, p_r] \rangle_{\hat{\text{AV}},0} = i_{\text{eff}} \hbar - \tau \langle \dot{q}_r' \text{Tr}\tilde{C} p_r' \rangle_{\hat{\text{AV}},0} \quad r \text{ bosonic}
$$
$$
\langle \{q_r, p_r\} \rangle_{\hat{\text{AV}},0} = i_{\text{eff}} \hbar - \tau \langle \dot{q}_r' \text{Tr}\tilde{C} p_r' \rangle_{\hat{\text{AV}},0} \quad r \text{ fermionic}. \tag{5.33a}
$$

These are exact expressions, in which the terms proportional to τ are very small corrections when we can assume support properties analogous to those stated in assumption (1) of Section 5.3, but without the "eff" projections. The relations of Eq. (5.33a) correspond to taking x_u as a canonical coordinate q and x_t as a canonical momentum p; a similar set of relations is obtained by taking x_u as a p and x_t as a q. Equating the two expressions for the canonical commutators (anticommutators) obtained this way gives identities that are constraints on the averages over the canonical ensemble that appear in the τ terms; another way of getting these constraints is to note that the left-hand sides of the formulas in Eq. (5.33a) are antisymmetric (symmetric) under interchange of q_r with p_r, whereas the right-hand sides are not.

Substituting Eq. (5.33a) into the ensemble average of the expression of Eq. (5.1) for $\tilde{C}$, and using Eq. (4.11b), we get

$$
i_{\text{eff}}\hbar = \langle \tilde{C} \rangle_{\hat{A}V,0} = \langle \sum_{r \in B}[q_r, p_r] - \sum_{r \in F}\{q_r, p_r\} \rangle_{\hat{A}V,0}
$$

$$
= \left(\sum_{r \in B} - \sum_{r \in F} \right) i_{\text{eff}}\hbar - \tau \left(\sum_{r \in B} - \sum_{r \in F} \right) \langle \dot{q}_r' \operatorname{Tr}\tilde{C} p_r' \rangle_{\hat{A}V,0}.
$$

(5.33b)

After division by $\hbar$, transposition of terms, and use of $\Sigma_{r \in B}1 = n_B$, $\Sigma_{r \in F}1 = n_F$, this gives

$$
\left(\sum_{r \in B} - \sum_{r \in F} \right) \hbar^{-1}\tau \langle \dot{q}_r' \operatorname{Tr}\tilde{C} p_r' \rangle_{\hat{A}V,0} = i_{\text{eff}}(n_B - n_F - 1).
$$

(5.33c)

When the condition of approximation (1) of Section 5.3 is satisfied, the left-hand side of Eq. (5.33c) is a sum of very small terms. Let us consider the case in which r includes the spatial label of a translation invariant field theory, with sufficient convergence properties so that we may assume that this sum yields at most a finite, bounded result. Then the numbers of bosonic and fermionic degrees of freedom per unit volume contributing on the right-hand side of Eq. (5.33c) must be equal, since, if not, the right-hand side of Eq. (5.33c) would become infinite as the spatial volume grows to infinity, contradicting the boundedness of the left-hand side. Therefore, a trace dynamics that is a candidate pre-quantum mechanics must have equal numbers of bosonic and fermionic degrees of freedom (up to a finite residue). This is a much weaker requirement than supersymmetry, but of course is always satisfied by supersymmetric theories. When the numbers of bosonic and fermionic degrees of freedom are in balance, Eq. (5.33c) simplifies to

$$
\left(\sum_{r \in B} - \sum_{r \in F} \right) \hbar^{-1}\tau \langle \dot{q}_r' \operatorname{Tr}\tilde{C} p_r' \rangle_{\hat{A}V,0} = -i_{\text{eff}},
$$

(5.33d)

showing that the $\tau \dot{x}_r$ terms neglected in making approximation (1) sum in Eq. (5.33d) to give a total of unit magnitude.

The above analysis of boson–fermion balance has implications for the behavior of $\tilde{C}$ in the thermodynamic limit of large system size. Although the bosonic and fermionic contributions to $\tilde{C}$ each grow linearly with the size of the system, the near cancellation of their contributions to $\tilde{C}$ suggests that the total rate of growth of $\tilde{C}$ could be much smaller than those of the bosonic or fermionic parts taken separately. Hence, even though $\tilde{C}$ is formally an extensive thermodynamic quantity (it is additive for disjoint subsystems), it may remain bounded, or have much smaller than a linear rate of growth, as the system size gets large. In the first case $\hbar$ would remain bounded in the limit of large system size, while in the second case $\hbar$ could still be a weakly increasing function of system size; our general analysis does not determine the expected behavior. The behavior that we ascribe to $\tilde{C}$ is somewhat analogous to the behavior of the electric charge operator in electrically neutral matter: the summed charges Q_+ and Q_- of the positively charged and negatively charged constituents behave as extensive quantities, but they cancel in their contribution to the total charge.

5.5 Derivation of the Schrödinger equation

In Section 5.3 we have argued that the statistical thermodynamics of matrix models with a global unitary invariance leads to an emergent Heisenberg picture quantum mechanics. In this section we shall make the transition to the corresponding Schrödinger picture formulation. We take as our starting point the correspondence of operators in the underlying trace dynamics to operators in an effective quantum theory given in Eq. (5.23b). Transcribing the canonical commutators inside averages of Eqs. (5.21a–c) into operator statements, in the sense of the correspondence established in Section 5.3, we get

$$[X_{u\text{eff}}, \sigma_v X_{v\text{eff}}] = i_{\text{eff}}\hbar\omega_{uv}\sigma_v, \tag{5.34a}$$

which encodes the canonical commutators

$$[Q_{u\text{eff}}, Q_{v\text{eff}}] = [P_{u\text{eff}}, P_{v\text{eff}}] = 0, \quad [Q_{u\text{eff}}, P_{v\text{eff}}] = i_{\text{eff}}\hbar\delta_{uv} \tag{5.34b}$$

for u, v bosonic, and the effective canonical anticommutators

$$\{Q_{u\text{eff}}, Q_{v\text{eff}}\} = \{P_{u\text{eff}}, P_{v\text{eff}}\} = 0, \quad \{Q_{u\text{eff}}, P_{v\text{eff}}\} = i_{\text{eff}}\hbar\delta_{uv} \tag{5.34c}$$

for u, v fermionic, with all boson–fermion commutators vanishing. Similarly, the operator transcription of Eq. (5.19b) (in average) for the time evolution of $x_{u\text{eff}}$ becomes

$$\dot{X}_{u\text{eff}} = i_{\text{eff}}\hbar^{-1}[H_{\text{eff}}, X_{u\text{eff}}], \tag{5.34d}$$

which extends by the chain rule to

$$\dot{S}_{\text{eff}} = i_{\text{eff}}\hbar^{-1}[H_{\text{eff}}, S_{\text{eff}}], \tag{5.34e}$$

with S_{eff} any polynomial function of the operators $\{X_{r\text{eff}}\}$.

Before turning to the transition to the Schrödinger picture, we first discuss consistency issues raised by treating the fermionic anticommutators as operator equations. If we were to assign fermionic adjoint properties according to

$$Q_{r\text{eff}} = \Psi_{r\text{eff}}, \quad P_{r\text{eff}} = i\Psi^\dagger_{r\text{eff}}, \tag{5.35a}$$

then the non-vanishing anticommutator in Eq. (5.34c) would take the form

$$\{\Psi_{u\text{eff}}, \Psi^\dagger_{v\text{eff}}\} = -i i_{\text{eff}}\hbar\delta_{uv}. \tag{5.35b}$$

In the K-dimensional subspace of Hilbert space on which i_{eff} acts as $\pm i$, Eq. (5.35b) becomes

$$\{\Psi_{u\text{eff}}, \Psi^\dagger_{v\text{eff}}\} = \pm\hbar\delta_{uv}. \tag{5.35c}$$

The $+$ sign case of Eq. (5.35c) corresponds to the normal field theoretic fermionic anticommutator, but the $-$ sign case is inconsistent: Setting $u = v$, the $-$ sign case gives

$$\{\Psi_{v\text{eff}}, \Psi^\dagger_{v\text{eff}}\} = -\hbar, \tag{5.35d}$$

which is not possible because the left-hand side of this relation is the sum of two positive semidefinite operators. Therefore the correspondence of Eq. (5.23b) must include an extra $-$ sign in $\Psi^\dagger_{v\text{eff}}$ in the $i_{\text{eff}} = -i$ sector; in other words, a consistent form of the operator transcription of the adjointness assignment for fermions is given by

$$\psi_{r\text{eff}} = q_{r\text{eff}} \leftrightarrow Q_{r\text{eff}} = \Psi_{r\text{eff}}, \quad \psi^\dagger_{r\text{eff}} = p_{r\text{eff}} \leftrightarrow P_{r\text{eff}} = i\tau_3\Psi^\dagger_{r\text{eff}} = i_{\text{eff}}\Psi^\dagger_{r\text{eff}}, \tag{5.36a}$$

with τ_3 the 2×2 matrix introduced in Eq. (4.40). Correspondingly, the non-vanishing anticommutator in Eq. (5.35b) is changed to

$$\{\Psi_{u\text{eff}}, \Psi^\dagger_{v\text{eff}}\} = \hbar\delta_{uv}, \tag{5.36b}$$

which has the correct positive sign on the right in both the $i_{\text{eff}} = \pm i$ sectors of Hilbert space.

Turning to the boson sector, let us now introduce effective bosonic creation and annihilation operators in the effective quantum theory, denoted by $A_{r\text{eff}}$ and $A^\dagger_{r\text{eff}}$, by writing

$$Q_{r\text{eff}} = \frac{1}{\sqrt{2}}(A_{r\text{eff}} + A^\dagger_{r\text{eff}}), \quad P_{r\text{eff}} = \frac{1}{i_{\text{eff}}\sqrt{2}}(A_{r\text{eff}} - A^\dagger_{r\text{eff}}). \tag{5.37a}$$

Since i_{eff} commutes with all effective operators, the definition of Eq. (5.37a) is clearly consistent with the self-adjointness of Q_r and P_r. Rewriting the commutator algebra of Eq. (5.34b) in terms of $A_{r\text{eff}}$ and its adjoint, we get

$$[A_{u\text{eff}}, A_{v\text{eff}}] = [A^\dagger_{u\text{eff}}, A^\dagger_{v\text{eff}}] = 0, \quad [A_{u\text{eff}}, A^\dagger_{v\text{eff}}] = \hbar\delta_{uv}, \tag{5.37b}$$

which, as was the case in Eq. (5.36b), has the correct positive sign on the right in both the $i_{\mathrm{eff}} = \pm i$ sectors of Hilbert space. Note that again a factor of i_{eff} appears in the transformation from $Q_{r\mathrm{eff}}$, $P_{r\mathrm{eff}}$ to the corresponding creation and annihilation operators (which for both fermions and bosons, as defined above, differ by a factor of $\hbar^{\frac{1}{2}}$ from the customary ones).

We are now ready to discuss the transition from our emergent Heisenberg picture quantum mechanics to the Schrödinger picture, and to derive the usual nonrelativistic Schrödinger equation. Since the Heisenberg equation of motion of Eq. (5.34e) and the commutation relations in the form given in Eqs. (5.36b) and (5.37b) have the standard quantum mechanics form, what we do now is standard quantum theory, and makes no explicit reference to emergent origins of the quantum equations. Restricting ourselves to the case in which the effective Hamiltonian has no intrinsic time dependence, we define $U_{\mathrm{eff}}(t)$ by

$$U_{\mathrm{eff}}(t) = \exp(-i_{\mathrm{eff}}\hbar^{-1}t H_{\mathrm{eff}}), \qquad (5.38a)$$

so that

$$\frac{d}{dt}U_{\mathrm{eff}}(t) = -i_{\mathrm{eff}}\hbar^{-1} H_{\mathrm{eff}}U_{\mathrm{eff}}(t),$$

$$\frac{d}{dt}U_{\mathrm{eff}}(t)^{\dagger} = i_{\mathrm{eff}}\hbar^{-1} U_{\mathrm{eff}}(t)^{\dagger} H_{\mathrm{eff}}. \qquad (5.38b)$$

Then from the time-independent Heisenberg picture state vector ψ and a Heisenberg picture operator $S_{\mathrm{eff}}(t)$ with no intrinsic time dependence, we can form a Schrödinger picture state vector ψ_{Schr} and a Schrödinger picture operator $S_{\mathrm{eff\,Schr}}$ by the usual construction

$$\psi_{\mathrm{Schr}}(t) = U_{\mathrm{eff}}(t)\psi,$$

$$S_{\mathrm{eff\,Schr}} = U_{\mathrm{eff}}(t)S_{\mathrm{eff}}(t)U_{\mathrm{eff}}(t)^{\dagger}, \qquad (5.39a)$$

giving

$$i_{\mathrm{eff}}\hbar\frac{d}{dt}\psi_{\mathrm{Schr}}(t) = H_{\mathrm{eff}}\psi_{\mathrm{Schr}}(t),$$

$$\frac{d}{dt}S_{\mathrm{eff\,Schr}} = 0. \qquad (5.39b)$$

To derive the nonrelativistic Schrödinger equation, let us consider the spacetime continuum case in which r is the label $\vec{x}$, so that the fermionic anticommutation relations of Eq. (5.36b) take the form

$$\{\Psi_{\mathrm{eff}}(\vec{x}), \Psi^{\dagger}_{\mathrm{eff}}(\vec{y})\} = \hbar\delta^{3}(\vec{x} - \vec{y}). \qquad (5.40a)$$

For simplicity, we shall restrict our discussion to single particle wave functions, because no issues of principle are involved in the extension to the many-particle case. Since we are assuming that H_{eff} is bounded below and has the vacuum state as its lowest energy eigenstate, and since H_{eff} is expected to contain a standard quadratic kinetic term, conventional ladder operator arguments suggest that the nonrelativistic operator Ψ_{eff} should annihilate the vacuum state $|\text{vac}\rangle$. Assuming this, we have

$$\Psi_{\text{eff}}(\vec{x})|\text{vac}\rangle = 0, \tag{5.40b}$$

and sandwiching Eq. (5.40a) between $\langle\text{vac}|$ and $|\text{vac}\rangle$ we obtain

$$\langle\text{vac}|\Psi_{\text{eff}}(\vec{x})\Psi_{\text{eff}}^\dagger(\vec{y})|\text{vac}\rangle = \hbar\delta^3(\vec{x} - \vec{y}). \tag{5.40c}$$

In the bosonic case, we start from the bosonic commutation relation of Eq. (5.37b), and assuming that the nonrelativistic operator A_{eff} annihilates $|\text{vac}\rangle$, we end up with

$$\langle\text{vac}|A_{\text{eff}}(\vec{x})A_{\text{eff}}^\dagger(\vec{y})|\text{vac}\rangle = \hbar\delta^3(\vec{x} - \vec{y}), \tag{5.40d}$$

which has the same form as in the fermionic case. So it suffices to restrict ourselves henceforth to the fermionic case.

Let us now introduce a complete set of single fermion intermediate states into Eq. (5.40c), by inserting $1 = \sum_n |n\rangle\langle n|$ (with a discrete or continuous sum), giving

$$\sum_n \langle\text{vac}|\Psi_{\text{eff}}(\vec{x})|n\rangle\langle n|\Psi_{\text{eff}}^\dagger(\vec{y})|\text{vac}\rangle = \hbar\delta^3(\vec{x} - \vec{y}). \tag{5.41a}$$

If we now define a wave function $\Psi_n(\vec{x})$ by

$$\hbar^{\frac{1}{2}}\Psi_n(\vec{x}) = \langle\text{vac}|\Psi_{\text{eff}}(\vec{x})|n\rangle,$$
$$\hbar^{\frac{1}{2}}\Psi_n^*(\vec{x}) = \langle n|\Psi_{\text{eff}}^\dagger(\vec{x})|\text{vac}\rangle, \tag{5.41b}$$

then after dividing by $\hbar$, Eq. (5.41a) can be rewritten as

$$\sum_n \Psi_n(\vec{x})\Psi_n^*(\vec{y}) = \delta^3(\vec{x} - \vec{y}), \tag{5.41c}$$

which is the usual completeness relation in coordinate representation. Multiplying by $\int d^3y\,\Psi_m(\vec{y})$, we get

$$\sum_n \Psi_n(\vec{x})\int d^3y\,\Psi_n^*(\vec{y})\Psi_m(\vec{y}) = \Psi_m(\vec{x}), \tag{5.41d}$$

which by linear independence of the Ψ_n tells us that

$$\int d^3y\,\Psi_n^*(\vec{y})\Psi_m(\vec{y}) = \delta_{nm}, \tag{5.41e}$$

which is the orthonormality condition in coordinate representation. Taking the time derivative of the first line of Eq. (5.41b) and using the Heisenberg equation of motion of Eq. (5.34d), we get

$$\hbar^{\frac{1}{2}}\frac{d}{dt}\Psi_n(\vec{x}) = \langle \text{vac}|\frac{d}{dt}\Psi_{\text{eff}}(\vec{x})|n\rangle$$

$$= \langle \text{vac}|i_{\text{eff}}\hbar^{-1}[H_{\text{eff}}, \Psi_{\text{eff}}(\vec{x})]|n\rangle. \tag{5.42a}$$

If we take H_{eff} to be a one body operator of the form

$$H_{\text{eff}} = \int d^3y \Psi_{\text{eff}}^{\dagger}(\vec{y})\mathcal{H}_{\text{eff}}(\vec{y})\Psi_{\text{eff}}(\vec{y}), \tag{5.42b}$$

then the commutator appearing in Eq. (5.42a) is given by

$$[H_{\text{eff}}, \Psi_{\text{eff}}(\vec{x})] = -\mathcal{H}_{\text{eff}}(\vec{x})\Psi_{\text{eff}}(\vec{x}), \tag{5.42c}$$

and so the right-hand side of Eq. (5.42a) becomes

$$-i_{\text{eff}}\hbar^{-1}\mathcal{H}_{\text{eff}}(\vec{x})\langle \text{vac}|\Psi_{\text{eff}}(\vec{x})|n\rangle = -i_{\text{eff}}\hbar^{-\frac{1}{2}}\mathcal{H}_{\text{eff}}(\vec{x})\Psi_n(\vec{x}). \tag{5.42d}$$

Multiplying through by $i_{\text{eff}}\hbar^{\frac{1}{2}}$, Eq. (5.42a) then yields the standard nonrelativistic Schrödinger equation in coordinate representation

$$i_{\text{eff}}\hbar\frac{d}{dt}\Psi_n(\vec{x}) = \mathcal{H}_{\text{eff}}(\vec{x})\Psi_n(\vec{x}). \tag{5.43}$$

This is of course all standard quantum mechanics and quantum field theory (which is why its extension to many-body wave functions and Fock space poses no difficulties). The point of going through it in detail is to emphasize that once we have obtained emergent canonical commutation relations and an emergent Heisenberg equation of motion for operators, the Schrödinger picture and Schrödinger equation of quantum mechanics follow in a straightforward way. To complete the argument for an emergent quantum mechanics, we must address the issue of how the probability interpretation (the Born rule) follows from our framework; this is the topic of the next chapter. But before turning to this, we shall first discuss why the emergent quantum mechanics that we have derived from an underlying deterministic trace dynamics evades the standard Kochen–Specker theorem and Bell inequality arguments against "hidden variable" theories.

5.6 Evasion of the Kochen–Specker theorem and Bell inequality arguments

The idea that the probabilistic nature of quantum mechanics may reflect the existence of hidden variables that determine individual outcomes is an old one, and has been much discussed in the physics, mathematics, and philosophy of science literature. Two important barriers that stand in the way of a hidden variable extension of quantum theory are "no-go" theorems that have been proved by Kochen and Specker (1967), on the one hand, and by Bell (1964, 1987), on the other. These theorems show that under certain quite general, and apparently innocuous assumptions, hidden variable interpretations of quantum mechanics either lead to a mathematical contradiction, in the Kochen–Specker case, or to a conflict with experiment, in the Bell case. Our aim in this section is to briefly analyze the assumptions that enter into these "no-go" theorems, and to show why they are evaded in our construction of quantum theory as an emergent phenomenon in trace dynamics models.

We begin with the Kochen–Specker argument, which is based in its original form on special properties of the spin-1 angular momentum matrices. Letting S_k denote the spin operators with matrix elements $(S_k)_{lm} = i\epsilon_{klm}$, where ϵ_{klm} is the three index completely antisymmetric tensor (we have set $\hbar$ equal to 1), a simple calculation shows that for any three mutually orthogonal directions $1, 2, 3$, the operators S_1^2, S_2^2, S_3^2 form a mutually commuting set. Since these operators each have eigenvalues 0 or 1, and since $S_1^2 + S_2^2 + S_3^2 = 2$, in any measurement two of these three operators must take the value 1, and one must take the value 0. Let us now assume that there are hidden variables internal to the quantum mechanical spin system being measured, in one-to-one correspondence with the operators S_k^2, that take preassigned values which determine whether the measurement outcome is a 0 or a 1. Kochen and Specker show that this is impossible, by identifying a set of directions on the unit sphere that have mutually commuting spin-squared operators, but which admit no consistent assignment of 0s and 1s subject to the rule that each orthogonal triple must contain two 1s and one 0. (Although their original construction is very complicated, a simplified version of their argument has been given by Peres, 1993.) This shows that the quantum mechanics of spin 1 systems cannot be reproduced by a set of hidden variables that take precise values which determine the measurement outcomes.

In the derivation of emergent quantum mechanics given in Section 5.3, the assumptions of the Kochen–Specker argument are not fulfilled. In the precursor trace dynamics, the matrix hidden dynamical variables corresponding to a spin triple interact with all other dynamical variables, and in particular with those associated with the measuring apparatus used to measure the spin-squared component. Since different apparatus configurations are needed to measure the different spin-squared components, the Kochen–Specker assumption that the hidden variables are internal to just the measured spin system, and have no contextual dependence on the apparatus used to measure the spin, breaks down. Moreover, the assertion that the squares of the spin operators, when measured, take exclusively the values 0 and 1, is a property of the states of the emergent quantum theory. This assertion depends directly on the emergent canonical commutation relations, whereas the operator algebras in the underlying theory involve matrix variables with unspecified commutation properties. Thus, the Kochen–Specker argument cannot be formulated at the underlying trace dynamics level. Note that in talking about measurements, we are really going beyond what has been achieved so far in our construction of emergent quantum theory, since in Section 5.3 we derived the unitary dynamics and canonical algebra of quantum mechanics, but said nothing about how the wave function assigns probabilities to measurement outcomes. A detailed discussion of the emergence of probabilities will be given in Chapter 6; we shall take a quick look ahead at this discussion shortly, in the course of analyzing the breakdown of the Bell inequalities.

We consider next the Bell inequalities, which have engendered an immense and still growing literature. Consider a spin-0 state that decays into two spin-1/2 particles, which move in opposite directions towards left and right Stern–Gerlach analyzers A and B. The analyzers are constructed so that they measure the component of spin along axes at respective angles α, β with respect to the z-axis. Quantum mechanics assigns a probability $P(a, b|\alpha, \beta)$ for observing spin components $a = \pm 1/2$, $b = \pm 1/2$ in the left and right analyzers. Suppose now that there are hidden variables λ that determine the outcomes, so that there is a function $P(a, b|\alpha, \beta, \lambda)$ that takes only the values 0 or 1 on its support. If $\rho(\lambda)$ is the probability distribution of the hidden variables, then the law of conditional probability tells us that

$$P(a, b|\alpha, \beta) = \int d\lambda \rho(\lambda) P(a, b|\alpha, \beta, \lambda). \tag{5.44a}$$

Bell now adds the additional requirement of local causality, which states that the result from analyzer A must be independent of the setting of analyzer B, and vice versa, since the setting angles α and β can be chosen just before the measurements are made, not allowing time for a light signal to propagate across the distance separating the analyzers. Local causality requires that $P(a, b|\alpha, \beta, \lambda)$ must factorize according to

$$P(a, b|\alpha, \beta, \lambda) = P_A(a|\alpha, \lambda) P_B(b|\beta, \lambda), \tag{5.44b}$$

which when substituted into Eq. (5.44a) gives

$$P(a, b|\alpha, \beta) = \int d\lambda \rho(\lambda) P_A(a|\alpha, \lambda) P_B(b|\beta, \lambda). \tag{5.44c}$$

From Eq. (5.44c) Bell (1964), and subsequently Clauser *et al.* (1969), have derived inequalities on $P(a, b | \alpha, \beta)$ that are not satisfied by the quantum mechanical expression for this probability. For a two-photon analog of Eq. (5.44c), the quantum mechanical prediction, in the regime where the analogous inequalities are violated, has been verified in experiments of Aspect and others. Thus, apart from technical caveats that are still being debated, but that we believe are not decisive, one is forced to the conclusion that if there are hidden variables that determine quantum mechanical outcomes, they in some manner must violate the assumption of local causality.

Turning now to our derivation of an emergent quantum mechanics from trace dynamics, we see that if the usual Born probability interpretation also emerges as a feature of trace dynamics, then the Bell inequalities must be violated, just as they are in standard quantum mechanics. Hence Bell's argument tells us that a violation of local causality must be involved at some stage. Although the matrix dynamical variables of trace dynamics are non-commutative in a completely general way, this is not necessarily enough, since the Poincaré invariant models studied in Sections 2.3 and 2.4 and Chapter 3 have quadratic Lagrangian terms emulating the familiar field theory forms. They thus give rise to retarded Green's functions of the usual form which can be used to perturbatively integrate the equations forward in time, giving rise to the usual light cone causal constraints. Were this not so, it would be possible to send information at superluminal speeds, contradicting observation. However, our Ward identity derivation of Section 5.1 contains a source of violation of local causality, which is the way in which the operator $\tilde{C}$ enters into the final term in Eq. (5.14b). Since we have seen that the boson and fermion contributions to $\tilde{C}$ largely cancel, despite the fact that $\tilde{C}$ is an extensive quantity, it can have large fluctuations over the operator phase space. Correspondingly, in any finite subsystem of the universe described by the canonical ensemble, $\tilde{C}$ has large fluctuations over the ensemble and hence as a function of time. These fluctuations give rise to corrections to the emergent quantum mechanics that we derived by replacing $\tilde{C}$ by its ensemble average.

In Chapter 6 we argue that these fluctuations do not affect the unit normalization of states, but add stochastic terms to the effective Schrödinger equation that describes the time development of a state. In order to preserve state normalization, this Schrödinger equation in the generic case must be nonlinear in the state, which introduces violations of local causality, since changes in the wave function at one spatial point are instantaneously communicated, via the noise terms, to all spatial points. We shall show explicitly that, by making simple and plausible models for the structure of the fluctuations in $\tilde{C}$, they play the role of a Brownian "noise" which drives state vector reduction, in such a way as to be precisely consistent with Born rule probabilities. At the same time, we shall see that while the Brownian terms in the stochastic Schrödinger equation are nonlinear, the average over the noise of the density matrix obeys a linear evolution equation, making superluminal signal transmission impossible. Thus, since the probability interpretation of emergent quantum mechanics comes about through the intervention of nonlocal effects induced by the nonlinearity of the stochastic terms in the Schrödinger equation, the assumptions underlying the Bell analysis are not obeyed, and the emergent theory is not constrained by the Bell inequalities.

6

Brownian motion corrections to Schrödinger dynamics and the emergence of the probability interpretation

Up to this point we have worked in the thermodynamic limit, with our reasoning based on the study of averages of dynamical variables in the canonical ensemble, with all fine grained structure averaged out. However, as in classical statistical mechanics, there are contexts in which fluctuations around the averages, which can be modeled in a natural way by a generalized Brownian motion, are important. We shall argue in this chapter that Brownian motion corrections to emergent quantum mechanics can provide the mechanism responsible both for reduction of the state vector, and for the emergence of the Born and Lüders probability rules. In Section 6.1, we introduce a simple model for the leading fluctuation corrections to the thermodynamic limit, and from it show that, with suitable assumptions, one can derive the standard stochastic Schrödinger equation for objective state vector reduction. Depending on details of the model, the stochastic driving terms in this equation can couple to the total energy, to a local density such as the energy or particle number density, or to both the energy and a local density. In Section 6.2, we give the proof that when the stochastic driving terms involve a set of mutually commuting operators, this equation leads to state vector reduction with Born rule probabilities. Finally, in Sections 6.3, 6.4 and 6.5 we discuss phenomenological constraints on stochastic models for objective state vector reduction, placed by the twin requirements that the excellent agreement of quantum mechanical predictions with experiments must be maintained, while still allowing state vector reduction to occur in measurement situations when the measuring apparatus itself is treated quantum mechanically. We first derive, in Section 6.3, formulas governing the reduction rate, and then apply them, in Sections 6.4 and 6.5, to the energy-driven and localization reduction models respectively.

6.1 Scenarios leading to the localization and the energy-driven
stochastic Schrödinger equations

As the starting point for obtaining stochastic corrections to the Schrödinger equation, we shall return to the general Ward identity of Eqs. (5.15a–c), in which the source terms have been varied and then set equal to zero. We continue to make approximations (1) and (2) of Section 5.3, that is, we neglect the τ and η terms in Eq. (5.15a), but now we do not make the approximation of replacing $\tilde{C}$ by its average value, so that Eq. (5.15a) takes the form

$$\mathcal{D}x_{\text{ueff}} = i_{\text{eff}}[W_{\text{eff}}, x_{\text{ueff}}] + \sum_{s,\ell} \omega_{us}\epsilon_\ell \left(W_s^{R\ell} \frac{1}{2}\{\tilde{C}, i_{\text{eff}}\} W_s^{L\ell} \right)_{\text{eff}}, \tag{6.1a}$$

and Eq. (5.15c) states that this expression vanishes inside suitable canonical ensemble averages. (To recapitulate some terminology, i_{eff} is defined, together with $\hbar$, from the ensemble average of $\tilde{C}$ in Eq. (4.11b); W is a general bosonic polynomial in the matrix dynamical variables $\{x_r\}$; the effective or "eff" projection is defined in Eqs. (4.41a–d); the structure coefficients ω_{us} and the grading factor ϵ_ℓ are defined in Eqs. (1.16a,b) and the line following Eq. (1.10b), respectively; and, finally, the decomposition of W into left and right factors $W_s^{L\ell}$ and $W_s^{R\ell}$ with respect to variations of x_s is defined in Eqs. (5.10a,b).) We shall now proceed in two steps: first we study the implications of Eq. (6.1a) for the normalization and completeness of wave functions at a fixed time, and then we study its implications for the time development of wave functions, that is, for the Schrödinger equation derived in Eq. (5.43).

For the first step we observe, recalling the discussion preceding Eq. (5.31c), that if we take $W = \sigma_t x_t$, then there is only one term in the sum over ℓ, with $\epsilon_{\ell=1} = 1$, $W_s^{L\ell=1} = \sigma_t \delta_{st}$, and $W_s^{R\ell=1} = 1$, so that the final term in Eq. (6.1a) reduces to

$$\omega_{ut} \frac{1}{2}\{\tilde{C}, i_{\text{eff}}\}\sigma_t. \tag{6.1b}$$

Continuing to assume that unitary fixing does not appreciably change the average of $\tilde{C}$, so that $\langle \tilde{C} \rangle_{\hat{A}V,0} \simeq \langle \tilde{C} \rangle_{AV} = i_{\text{eff}}\hbar$, this term has zero-source ensemble average

$$-\omega_{ut}\hbar\sigma_t, \tag{6.1c}$$

and so setting the sources equal to zero in Eq. (5.15b), we get

$$\langle i_{\text{eff}}[\sigma_t x_{t\text{eff}}, x_{u\text{eff}}] - \omega_{ut}\hbar\sigma_t \rangle_{\hat{A}V,0} = 0. \tag{6.2a}$$

Multiplying Eq. (6.2a) on the left by $\psi_0^\dagger$ and on the right by ψ_0, and assuming the correspondence between canonical ensemble averages at a specified initial time and Wightman functions given in Eq. (5.23b), we see that

$$\langle \text{vac}|i_{\text{eff}}[\sigma_t X_{t\text{eff}}, X_{u\text{eff}}] - \omega_{ut}\hbar\sigma_t|\text{vac}\rangle = 0. \tag{6.2b}$$

Using Eq. (6.2b) in place of Eq. (5.34a) as the starting point for the analysis of Section 5.5, we learn that the vacuum expectation of Eq. (5.40a) is still given by

$$\langle \text{vac}|\{\Psi_{\text{eff}}(\vec{x}), \Psi_{\text{eff}}^\dagger(\vec{y})\}|\text{vac}\rangle = \hbar\delta^3(\vec{x} - \vec{y}), \tag{6.2c}$$

even when fluctuations of $\tilde{C}$ about its ensemble average are taken into account. (Equation (6.2c) is written for the fermion case; the same conclusion holds for the boson case as well.) Thus, the orthonormalization and completeness of wave functions at an initial time on which the canonical ensemble is formed, derived in Eqs. (5.41a–e), also do not make use of the approximation of replacing $\tilde{C}$ by its average value, as a result of the insensitivity of the vacuum expectation of the canonical algebra to fluctuations in $\tilde{C}$. Although Eq. (6.2c) has been derived for single fermion operators, similar relations hold in the case of multiparticle composite operators, as long as the compositeness scale is small enough for the composite system to be effectively treated as a point particle. To the extent that the time evolution is approximated by a unitary Heisenberg evolution, the orthonormality of states is preserved in time. In this chapter we shall be exploring the effects of deviations from Heisenberg evolution, and we will see that in the approximation to the dynamics that we use, state normalization is preserved in time, but states that are initially orthogonal do not remain so under time evolution.

For the second step of our argument, we take W in Eq. (6.1a) to be the operator Hamiltonian H, as we did in Eq. (5.19b) of Section 5.3, but we now take the fluctuations of $\tilde{C}$ about its ensemble average into account. We denote this fluctuating term by

$$\Delta\tilde{C} = \tilde{C} - \langle\tilde{C}\rangle_{\hat{A}\text{V},0} \simeq \tilde{C} - i_{\text{eff}}\hbar, \tag{6.3a}$$

so that the expression entering into Eq. (6.1a) takes the form

$$\frac{1}{2}\{\tilde{C}, i_{\text{eff}}\} = -\hbar + \frac{1}{2}\{\Delta\tilde{C}, i_{\text{eff}}\}. \tag{6.3b}$$

Let us now rewrite the fluctuation term on the right of Eq. (6.3b) as

$$\frac{1}{2}\{\Delta\tilde{C}, i_{\text{eff}}\} = -\hbar(\mathcal{K} + \mathcal{N}), \tag{6.3c}$$

with $\mathcal{K}$ a fluctuating c-number, and with $\mathcal{N}$ a fluctuating matrix, with the splitting defined so that under the correspondence of Eq. (5.23b), the quantum operator analog of $\mathcal{N}$ is normal ordered, that is, has all annihilation operators ordered to the right. Since we are primarily interested in the effect of the fluctuations on nonrelativistic matter, we shall ignore boson field contributions to $\mathcal{N}$. Because the quantum operator analog of $\mathcal{N}$ has fermion number zero, retaining only fermion field contributions implies that $\mathcal{N}$ contains no terms with only creation operators, and thus it annihilates the vacuum state. We shall assume that the fluctuations are small quantities, so that working to first order in the fluctuations, we can separately determine the effects of the $\mathcal{K}$ and $\mathcal{N}$ fluctuation terms, and then add them. Hence there are implicit error terms of the form $O(\mathcal{K}^2)$, $O(\mathcal{N}^2)$, and $O(\mathcal{K}\mathcal{N})$ in our formulas, which we shall not indicate explicitly. However, we shall see that once the linear fluctuation terms in the stochastic Schrödinger equation have been determined, general physical requirements will determine the structure of the implicit higher-order terms.

As discussed in Chapter 2, when we make the standard fermionic adjointness assignment, $\tilde{C}$ is anti-self-adjoint, which makes $\mathcal{K}$ real and $\mathcal{N}$ self-adjoint. We shall find that this leads to a structure for the stochastic Schrödinger equation that does not lead to state vector reduction; what will be needed to get state vector reduction is for $\mathcal{K}$ to have an imaginary part and/or for $\mathcal{N}$ to have an anti-self-adjoint part, both corresponding to the presence of a self-adjoint piece in $\tilde{C}$. We shall assume that the introduction of a self-adjoint piece in $\tilde{C}$ does not change the form of the canonical ensemble ρ, and that the ensemble expectation of the self-adjoint part of $\tilde{C}$ vanishes, so that the self-adjoint part contributes only to the fluctuation $\Delta\tilde{C}$, leaving our formulas relating i_{eff} to the ensemble average of $\tilde{C}$ unchanged. To see that these assumptions can be consistently realized, let us return to the model set up in Eqs. (2.17a–c) and (2.18a–c), where we found that a self-adjoint part of $\tilde{C}$ has the general form

$$\tilde{C}^{\text{sa}} = -\frac{1}{2}\sum_{r,s\in F}[q_s q_r^\dagger, A_{rs}]. \tag{6.4a}$$

If we take A_{rs} to commute with i_{eff}, then by Eq. (4.39b) and the cyclic identities we have

$$\text{Tr}\tilde{\lambda}\tilde{C}^{\text{sa}} = -\frac{1}{2}\sum_{r,s\in F}\text{Tr}q_s q_r^\dagger[A_{rs}, \tilde{\lambda}] = 0, \tag{6.4b}$$

in other words, we get a real valued canonical ensemble without splitting $\text{Tr}\tilde{\lambda}\tilde{C}$ into self-adjoint and anti-self-adjoint parts as in Eq. (4.13d). Similarly, we find that

$$\langle\tilde{C}^{\text{sa}}\rangle_{\text{AV}} = -\frac{1}{2}\sum_{r,s\in F}[\langle q_s q_r^\dagger\rangle_{\text{AV}}, A_{rs}] = 0, \tag{6.4c}$$

since the ensemble average $\langle q_s q_r^\dagger\rangle_{\text{AV}}$ is a function solely of i_{eff}. However, the assumption that A_{rs} commutes with i_{eff} does not imply the vanishing of $\{i_{\text{eff}}, \tilde{C}^{\text{sa}}\}$, which is given by

$$\{i_{\text{eff}}, \tilde{C}^{\text{sa}}\} = -\frac{1}{2} \sum_{r,s \in F} [\{i_{\text{eff}}, q_s q_r^\dagger\}, A_{rs}] \neq 0, \tag{6.4d}$$

and so there will be non-vanishing fluctuation contributions to $\mathcal{K}$ and $\mathcal{N}$, but otherwise the structure of the Ward identities, which follows from the general form of the canonical ensemble, is unaltered.

Therefore, making the assumptions just described, we shall continue to use the Ward identity of Eqs. (5.15a–c), and the consequences derived from it in Section 5.3. We begin by discussing the effects of the $\mathcal{K}$ term in Eqs. (6.3b,c), which can be done in a model independent way. Since $\mathcal{K}$ is a c-number, when the terms $-\hbar(1 + \mathcal{K})$ in Eqs. (6.3b,c) are substituted into Eq. (6.1a) we can still use Eq. (5.10b) to evaluate the sum over ℓ, just as we did in Eq. (5.18c), giving to leading order in $\mathcal{K}$

$$\mathcal{D}x_{u\text{eff}} = i_{\text{eff}}[W_{\text{eff}}, x_{u\text{eff}}] - \hbar(1 + \mathcal{K}) \sum_s \omega_{us} \left(\frac{\delta \mathbf{W}}{\delta x_s}\right)_{\text{eff}}. \tag{6.5a}$$

Taking W to be the operator Hamiltonian H, we now find that Eq. (5.19b) is replaced by

$$\mathcal{D}x_{u\text{eff}} = i_{\text{eff}}[H_{\text{eff}}, x_{u\text{eff}}] - \hbar(1 + \mathcal{K})\dot{x}_{u\text{eff}}, \tag{6.5b}$$

and since $\mathcal{K}$ is a c-number, to leading order in $\mathcal{K}$ we can replace $\hbar\dot{x}_{u\text{eff}}$ by $i_{\text{eff}}[H_{\text{eff}}, x_{u\text{eff}}]$ in the $\mathcal{K}$ term, giving

$$\mathcal{D}x_{u\text{eff}} = i_{\text{eff}}[H_{\text{eff}}, x_{u\text{eff}}] - \hbar\dot{x}_{u\text{eff}} - \mathcal{K}i_{\text{eff}}[H_{\text{eff}}, x_{u\text{eff}}]. \tag{6.5c}$$

This equation is still to be interpreted as an expression that vanishes when sandwiched between general polynomials $S_{L,R}$, and averaged over the zero-source canonical ensemble.

We turn next to the evaluation of the contribution of the matrix part $\mathcal{N}$ of the fluctuation $\Delta\tilde{C}$. To make a concrete statement about the contribution of this term, we must introduce assumptions about the structure of the Hamiltonian H. Since we will be primarily interested in the effects of fluctuation corrections on a measurement apparatus that consists of nonrelativistic matter, with a Hamiltonian that is dominated by its rest mass terms, we will take as a simplified model for the H describing the system and apparatus

$$H = \sum_r m_r \int d^3x \frac{1}{2} i [\psi_r^\dagger(\vec{x}), \psi_r(\vec{x})] + \text{constant}$$

$$= \sum_r \sum_\ell \frac{1}{2} i m_r [\psi_{r\ell}^\dagger, \psi_{r\ell}] + \text{constant}. \tag{6.6}$$

Here m_r is the rest mass of the rth species, and we have first written the operator terms in H in antisymmetrized form in terms of a coordinate basis $\vec{x}$, and then written them in terms of an arbitrary basis ℓ labeling a general complete set. In writing Eq. (6.6) we have assumed that for the dynamical variables describing ordinary matter, the matrix A_{rs} appearing in Eq. (6.4a) is simply δ_{rs}, so that the standard adjointness assignment of Eq. (2.4b) applies; this does not rule out the presence of a nontrivial A_{rs} for degrees of freedom that are much higher energy than those involved in laboratory measurements. There is an additional assumption implicit in our use of the antisymmetrized form of the mass term in Eq. (6.6), arising from the fact that the symmetrized combination $i\{\psi_{r\ell}^{\dagger}, \psi_{r\ell}\}$ has vanishing trace and so does not contribute to the trace Hamiltonian **H**. Changing the operator-ordering prescription by adding such a term to the operator Hamiltonian H changes the contribution coming from the operator-valued fluctuation term $\mathcal{N}$ of Eq. (6.3c). We deal with this ambiguity by assuming that where factor ordering of the operator Hamiltonian makes a difference in the calculation of the fluctuation terms, the Weyl-ordered recipe, which as we have seen in Section 1.5 minimizes the differences between trace dynamics calculations and the corresponding canonical quantum mechanical ones, is the appropriate ordering convention.

Since the model for H in Eq. (6.6) is bilinear in the fields, it is now easy to evaluate the final term in Eq. (6.1a) when W is taken as H, and when x_u is taken first as $\psi_{r\ell}$ and then as $\psi_{r\ell}^{\dagger}$. Including the leading-order result of Eq. (6.5c) for the $\mathcal{K}$ contribution, together with the approximate result for the $\mathcal{N}$ contribution obtained by using the model of Eq. (6.6) for H and the definitions of Eqs. (5.10a) and (1.16a,b), we get

$$\mathcal{D}\psi_{r\ell\text{eff}} = i_{\text{eff}}[H_{\text{eff}}, \psi_{r\ell\text{eff}}] - \hbar\dot{\psi}_{r\ell\text{eff}} - \mathcal{K}i_{\text{eff}}[H_{\text{eff}}, \psi_{r\ell\text{eff}}] + \frac{1}{2}i\hbar m_r\{\mathcal{N}, \psi_{r\ell\text{eff}}\},$$

$$\mathcal{D}\psi_{r\ell\text{eff}}^{\dagger} = i_{\text{eff}}[H_{\text{eff}}, \psi_{r\ell\text{eff}}^{\dagger}] - \hbar\dot{\psi}_{r\ell\text{eff}}^{\dagger} - \mathcal{K}i_{\text{eff}}[H_{\text{eff}}, \psi_{r\ell\text{eff}}^{\dagger}] - \frac{1}{2}i\hbar m_r\{\mathcal{N}, \psi_{r\ell\text{eff}}^{\dagger}\}.$$

$$(6.7a)$$

These are again to be interpreted as expressions that vanish when sandwiched between general polynomials $S_{L,R}$, and averaged over the zero-source canonical ensemble. We now wish to attempt to reinterpret the vanishing of Eq. (6.7a) as an operator statement in the effective quantum field theory, restricting ourselves henceforth to the $i_{\text{eff}} = i$ sector. To do this several problems must be addressed, that require further assumptions.

Since $\mathcal{K}$ and $\mathcal{N}$ are rapidly fluctuating quantities in the underlying matrix phase space, they do not have a direct transcription to the effective field theory. However, consistent with the idealization involved in describing an ergodic, time-dependent matrix dynamics by the static canonical ensemble, it is natural to model them in the effective field theory transcription of Eq. (6.7a) as time-dependent Brownian motions $\mathcal{K}(t)$ and $\mathcal{N}(t)$. Hence we provisionally reinterpret Eqs. (6.7a) and (5.15c) as field theory equations of motion

$$\dot{\Psi}_{r\ell\text{eff}} = i\hbar^{-1}[H_{\text{eff}}, \Psi_{r\ell\text{eff}}] - \mathcal{K}(t)i\hbar^{-1}[H_{\text{eff}}, \Psi_{r\ell\text{eff}}] + \frac{1}{2}im_r\{\mathcal{N}(t), \Psi_{r\ell\text{eff}}\},$$

$$\dot{\Psi}^\dagger_{r\ell\text{eff}} = i\hbar^{-1}[H_{\text{eff}}, \Psi^\dagger_{r\ell\text{eff}}] - \mathcal{K}(t)i\hbar^{-1}[H_{\text{eff}}, \Psi^\dagger_{r\ell\text{eff}}] - \frac{1}{2}im_r\{\mathcal{N}(t), \Psi^\dagger_{r\ell\text{eff}}\}.$$

$$\tag{6.7b}$$

(The factor of i arising from the correspondence between $\psi^\dagger_{r\ell\text{eff}}$ and $\Psi^\dagger_{r\ell\text{eff}}$ is common to all terms of Eq. (6.7b), and so cancels out.) When $\tilde{C}$ is anti-self-adjoint, so that $\mathcal{K}(t)$ is real and $\mathcal{N}(t)$ is self-adjoint, the operator equations for $\dot{\Psi}_{r\ell\text{eff}}$ and $\dot{\Psi}^\dagger_{r\ell\text{eff}}$ are consistent with one another, in that the right-hand side of the equation for $\dot{\Psi}^\dagger_{r\ell\text{eff}}$ is the adjoint of the right-hand side of the equation for $\dot{\Psi}_{r\ell\text{eff}}$. However, when $\tilde{C}$ has a self-adjoint part, so that $\mathcal{K}(t)$ has an imaginary part and $\mathcal{N}(t)$ has an anti-self-adjoint part, the operator equations for $\dot{\Psi}_{r\ell\text{eff}}$ and $\dot{\Psi}^\dagger_{r\ell\text{eff}}$ are no longer consistent, since the adjoints of the contributions of the imaginary part of $\mathcal{K}(t)$ and the anti-self-adjoint part of $\mathcal{N}(t)$ to the right-hand side of the equation for $\dot{\Psi}_{r\ell\text{eff}}$ appear with the wrong sign to give the right-hand side of the equation for $\dot{\Psi}^\dagger_{r\ell\text{eff}}$. This inconsistency indicates that when the fluctuations in $\tilde{C}$ have a self-adjoint part, we cannot directly apply the correspondence postulated in Eq. (5.23a) between operators in the underlying trace dynamics and operators in the emergent effective field theory. We propose to avoid this inconsistency by regarding Eqs. (6.7b) not as operator equations, but rather as constraints on the vacuum state $|\text{vac}\rangle$. Since $\mathcal{N}(t)$, as well as $\Psi_{r\ell\text{eff}}$, and (with an appropriate choice of the constant in Eq. (6.6)) H_{eff} all annihilate the vacuum, all terms in the equation for $\dot{\Psi}_{r\ell\text{eff}}$ annihilate the vacuum, so this equation reduces to the triviality $0 = 0$, while the equation for $\dot{\Psi}^\dagger_{r\ell\text{eff}}$ simplifies to

$$\dot{\Psi}^\dagger_{r\ell\text{eff}}|\text{vac}\rangle = [i\hbar^{-1}H_{\text{eff}} - \mathcal{K}(t)i\hbar^{-1}H_{\text{eff}} - \frac{1}{2}im_r\mathcal{N}(t)]\Psi^\dagger_{r\ell\text{eff}}|\text{vac}\rangle. \tag{6.7c}$$

Splitting $\mathcal{K}$ into real and imaginary parts, and $\mathcal{N}$ into self-adjoint and anti-self-adjoint parts, which we label with the respective subscripts 0 and 1, Eq. (6.7c) takes the form

$$\dot{\Psi}^\dagger_{r\ell\text{eff}}|\text{vac}\rangle = [i\hbar^{-1}H_{\text{eff}} - \mathcal{K}_0(t)i\hbar^{-1}H_{\text{eff}} + \mathcal{K}_1(t)\hbar^{-1}H_{\text{eff}} - \frac{1}{2}im_r\mathcal{N}_0(t)$$

$$+ \frac{1}{2}m_r\mathcal{N}_1(t)]\Psi^\dagger_{r\ell\text{eff}}|\text{vac}\rangle. \tag{6.7d}$$

The appropriate Hilbert space basis for a measured system together with the measuring apparatus is provided by a set of states $|\Phi(\{r, \ell\})\rangle$ obtained by acting on the vacuum with a product of many different creation operators $\Psi^\dagger_{r\ell\text{eff}}$ for different values of r and ℓ. To infer the equation of motion for $|\Phi(\{r, \ell\})\rangle$ that corresponds to Eq. (6.7d), let us adopt for the moment an independent particle picture, so that the operators H_{eff}, $\mathcal{N}_0(t)$ and $\mathcal{N}_1(t)$ are sums of operators, that each act on individual fermion degrees of freedom labeled by r and ℓ, and that commute with all other fermion degrees of freedom. This assumption is clearly consistent with the model for the dominant term in the Hamiltonian given in Eq. (6.6). So we write

$$
\begin{aligned}
H_{\text{eff}} &= \sum_{r\ell} H_{\text{eff};r\ell}, \\
\mathcal{N}_0(t) &= \sum_{r\ell} \mathcal{N}_0(t)_{r\ell}, \\
\mathcal{N}_1(t) &= \sum_{r\ell} \mathcal{N}_1(t)_{r\ell},
\end{aligned}
\tag{6.8a}
$$

with all terms in the sums annihilating the vacuum state. Substituting Eq. (6.8a) into Eq. (6.7c), the latter becomes

$$
\begin{aligned}
\dot{\Psi}^\dagger_{r\ell\text{eff}}|\text{vac}\rangle = & [i\hbar^{-1}H_{\text{eff};r\ell} - \mathcal{K}_0(t)i\hbar^{-1}H_{\text{eff};r\ell} + \mathcal{K}_1(t)\hbar^{-1}H_{\text{eff};r\ell} \\
& - \frac{1}{2}im_r\mathcal{N}_0(t)_{r\ell} + \frac{1}{2}m_r\mathcal{N}_1(t)_{r\ell}]\Psi^\dagger_{r\ell\text{eff}}|\text{vac}\rangle.
\end{aligned}
\tag{6.8b}
$$

From this we see that a general basis state

$$
|\Phi(\{r, \ell\})\rangle = \prod_{r,\ell} \Psi^\dagger_{r\ell\text{eff}}|\text{vac}\rangle.
\tag{6.8c}
$$

obtained by acting on the vacuum with a product of creation operators $\Psi^\dagger_{r\ell\text{eff}}$ with different values of r and ℓ, will obey the equation of motion

$$
\begin{aligned}
|\dot{\Phi}(\{r, \ell\})\rangle = & [i\hbar^{-1}H_{\text{eff}} - \mathcal{K}_0(t)i\hbar^{-1}H_{\text{eff}} + \mathcal{K}_1(t)\hbar^{-1}H_{\text{eff}} - \frac{1}{2}i\mathcal{M}_0(t) \\
& + \frac{1}{2}\mathcal{M}_1(t)]|\Phi(\{r, \ell\})\rangle,
\end{aligned}
\tag{6.8d}
$$

where we have introduced the definition

$$
\mathcal{M}_{0,1}(t) = \sum_{r\ell} m_r\mathcal{N}_{0,1}(t)_{r\ell}.
\tag{6.8e}
$$

Generalizing beyond the independent particle picture, we shall take Eqs. (6.8d,e) as the equation of motion of a generic many particle basis state. Up to this point we have stayed in Heisenberg picture, in which the basis states carry the time dependence and the state $|\Phi\rangle$ representing the physical state of the measured system and measuring apparatus is fixed in time. The time evolution of the wave function $\langle\Phi(\{r, \ell\})|\Phi\rangle$ will be the same, however, if we go over to the Schrödinger picture in which the basis states $|\Phi(\{r, \ell\})\rangle$ are constant in time, and in which the physical state $|\Phi\rangle$ carries the time dependence by evolving in the opposite sense to Eq. (6.8d), so that

$$|\dot{\Phi}\rangle = [-i\hbar^{-1}H_{\text{eff}} + \mathcal{K}_0(t)i\hbar^{-1}H_{\text{eff}} - \mathcal{K}_1(t)\hbar^{-1}H_{\text{eff}} + \frac{1}{2}i\mathcal{M}_0(t) - \frac{1}{2}\mathcal{M}_1(t)]|\Phi\rangle.$$

$$(6.8f)$$

When $\mathcal{K}_1$ and $\mathcal{M}_1$ are non-vanishing, the modified Schrödinger equation of Eq. (6.8f) does not preserve the norm of the state $|\Phi\rangle$, which we suggest is a reflection of the approximation involved in transcribing the time independent $\Delta\tilde{C}$ that fluctuates over the underlying matrix phase space into time-fluctuating processes $\mathcal{K}(t)$ and $\mathcal{N}(t)$ in the effective field theory. Since in the nonrelativistic limit all particle species are conserved in number, we must restore conservation of the norm under time evolution in order to obtain the physical state vector $|\Psi\rangle$. A norm-preserving Schrödinger equation, incorporating the fluctuation corrections, can be obtained by identifying the physical state $|\Psi\rangle$ with the renormalized $|\Phi\rangle$

$$|\Psi\rangle = \frac{|\Phi\rangle}{\langle\Phi|\Phi\rangle^{\frac{1}{2}}}. \qquad (6.9)$$

To complete the specification of the stochastic dynamics for $|\Psi\rangle$, we must specify the nonvanishing time averages and the operator structure of the fluctuating terms in Eqs. (6.8d,f). We shall make the assumption that the fluctuations can be described as linear superpositions of white noise terms; this seems reasonable both because we are postulating a large hierarchy of magnitude between the length scale characterizing the fluctuations of $\tilde{C}$ and the length scale characterizing the emergent quantum degrees of freedom and because in standard statistical mechanics applications of Brownian motion, the assumption of a white noise spectrum is generally a good approximation. Once we are dealing with white noise, the calculations are made much more tractable by using the standard Itô calculus representation of Brownian motion (for a pedagogical introduction to the Itô calculus, see Gardiner (1990)), in which each independent Brownian motion is represented by a differential dW_t^n, which together with dt obey the Itô calculus rules

$$(dW_t^n)^2 = \gamma_n dt,$$
$$dW_t^n dW_t^m = 0, m \neq n,$$
$$dW_t^n dt = dt^2 = 0. \qquad (6.10a)$$

For the c-number fluctuations $\mathcal{K}_{0,1}$ we write

$$i\hbar^{-1}\mathcal{K}_0(t)dt = i\beta_I dW_t^I,$$
$$-\hbar^{-1}\mathcal{K}_1(t) = \beta_R dW_t^R, \qquad (6.10b)$$

with

$$(dW_t^R)^2 = (dW_t^I)^2 = dt, \quad dW_t^R dW_t^I = 0. \tag{6.10c}$$

We allow for the possibility that the operator fluctuations $\mathcal{M}_{0,1}$ include a spatially correlated noise structure, by writing

$$\frac{1}{2} i \mathcal{M}_0(t) dt = i \int d^3x \, dW_t^I(\vec{x}) \mathcal{M}^I(\vec{x}),$$

$$-\frac{1}{2} \mathcal{M}_1(t) = \int d^3x \, dW_t^R(\vec{x}) \mathcal{M}^R(\vec{x}), \tag{6.10d}$$

with

$$dW_t^I(\vec{x}) dW_t^I(\vec{y}) = \gamma dt \delta^3(\vec{x} - \vec{y}),$$

$$dW_t^R(\vec{x}) dW_t^R(\vec{y}) = \gamma dt \delta^3(\vec{x} - \vec{y}),$$

$$dW_t^I(\vec{x}) dW_t^R(\vec{y}) = 0,$$

$$dW_t^I dW_t^I(\vec{x}) = dW_t^I dW_t^R(\vec{x}) = 0,$$

$$dW_t^R dW_t^I(\vec{x}) = dW_t^R dW_t^R(\vec{x}) = 0. \tag{6.10e}$$

Thus Eq. (6.8f) now takes the form

$$|d\Phi\rangle = [-i\hbar^{-1} H_{\text{eff}} dt + i\beta_I dW_t^I H_{\text{eff}} + \beta_R dW_t^R H_{\text{eff}}$$

$$+ i \int d^3x \, dW_t^I(\vec{x}) \mathcal{M}^I(\vec{x}) + \int d^3x \, dW_t^R(\vec{x}) \mathcal{M}^R(\vec{x})] |\Phi\rangle, \tag{6.10f}$$

up to implicit quadratic terms in the noise variables.

Equation (6.10f) is our basic result for the fluctuation modified Schrödinger equation, in the linearized approximation. In essence, what we have found is that the fluctuations of $\tilde{C}$ have the effect of replacing the usual Schrödinger equation by a stochastic Schrödinger equation containing Brownian motion terms. Although we have only retained linear terms in the fluctuations in Eq. (6.10f), we shall now show that in the white noise case, the quadratic terms are completely determined by imposing the two general structural requirements of norm preservation and the vanishing of superluminal signaling effects. Since according to the Itô calculus rules, there are no cubic and higher-order fluctuation terms, this completely specifies the structure of the stochastic Schrödinger equation.

To calculate the stochastic Schrödinger equation obeyed by the normalized state vector $|\Psi\rangle$, we must evaluate the differential of Eq. (6.9), taking care to use the Itô product rule (or stochastic integration by parts formula)

$$d(FG) = (dF)G + F(dG) + dFdG, \tag{6.11a}$$

which is the extension to the Itô calculus of the usual Leibniz chain rule for differentiation. Defining

$$\delta = \frac{\langle d\Phi|\Phi\rangle + \langle\Phi|d\Phi\rangle + \langle d\Phi|d\Phi\rangle}{\langle\Phi|\Phi\rangle}, \tag{6.11b}$$

we readily find

$$|d\Psi\rangle = \frac{|\Phi\rangle + |d\Phi\rangle}{\langle\Phi|\Phi\rangle^{\frac{1}{2}}(1+\delta)^{\frac{1}{2}}} - |\Psi\rangle$$

$$= |\Psi\rangle\left(-\frac{1}{2}\delta + \frac{3}{8}\delta^2\right) + \frac{|d\Phi\rangle}{\langle\Phi|\Phi\rangle^{\frac{1}{2}}}\left(1 - \frac{1}{2}\delta\right), \tag{6.11c}$$

with $|d\Phi\rangle$ given by Eq. (6.10f). From Eq. (6.10f), we calculate δ to be

$$\delta = 2\beta_R dW_t^R\langle H_{\text{eff}}\rangle + 2\int d^3x\, dW_t^R(\vec{x})\langle\mathcal{M}^R(\vec{x})\rangle$$

$$+ (\beta_R^2 + \beta_I^2)dt\langle H_{\text{eff}}^2\rangle + \gamma dt\int d^3x\langle\mathcal{M}^I(\vec{x})^2 + \mathcal{M}^R(\vec{x})^2\rangle, \tag{6.11d}$$

with the expectation $\langle\ \rangle$ defined by

$$\langle\mathcal{O}\rangle = \frac{\langle\Phi|\mathcal{O}|\Phi\rangle}{\langle\Phi|\Phi\rangle} = \langle\Psi|\mathcal{O}|\Psi\rangle. \tag{6.11e}$$

Writing Eq. (6.10f) as $|d\Phi\rangle = G|\Phi\rangle$ and substituting it and Eq. (6.11d) into Eq. (6.11c), we get the following result for $|d\Psi\rangle$

$$|d\Psi\rangle = [-\frac{1}{2}\delta + \frac{3}{8}\delta^2 + \left(1 - \frac{1}{2}\delta\right)G]|\Psi\rangle$$

$$= [X + Qdt + Rdt - \langle Q + R\rangle dt]|\Psi\rangle, \tag{6.12a}$$

with the operator X given by

$$X = -i\hbar^{-1}H_{\text{eff}}dt - \frac{1}{2}\beta_I^2 H_{\text{eff}}^2 dt - \frac{1}{2}\beta_R^2(H_{\text{eff}} - \langle H_{\text{eff}}\rangle)^2 dt$$

$$- \frac{1}{2}\gamma dt\int d^3x\,\mathcal{M}^I(\vec{x})^2 - \frac{1}{2}\gamma dt\int d^3x(\mathcal{M}^R(\vec{x}) - \langle\mathcal{M}^R(\vec{x})\rangle)^2$$

$$+ i\beta_I dW_t^I H_{\text{eff}} + \beta_R dW_t^R(H_{\text{eff}} - \langle H_{\text{eff}}\rangle)$$

$$+ i\int d^3x\, dW_t^I(\vec{x})\mathcal{M}^I(\vec{x}) + \int d^3x\, dW_t^R(\vec{x})(\mathcal{M}^R(\vec{x}) - \langle\mathcal{M}^R(\vec{x})\rangle)), \tag{6.12b}$$

with the operator Q given by

$$Q = \frac{1}{2}\beta_I^2 H_{\text{eff}}^2 + \frac{1}{2}\gamma \int d^3x \mathcal{M}^I(\vec{x})^2$$

$$+ \frac{1}{2}\beta_R^2(H_{\text{eff}}^2 - 4H_{\text{eff}}\langle H_{\text{eff}}\rangle) + \frac{1}{2}\int d^3x(\mathcal{M}^R(\vec{x})^2 - 4\mathcal{M}^R(\vec{x})\langle\mathcal{M}^R(\vec{x})\rangle),$$

$$(6.12c)$$

and with the operator R denoting the implicit quadratic terms in the noise that have been carried along through the calculation. The fact that R must enter Eq. (6.12a) through the combination $R - \langle R\rangle$ follows, as shown by Adler and Brun (2001), from the fact that since $|\Psi\rangle$ is normalized, we must have $\langle d\Psi|\Psi\rangle + \langle\Psi|d\Psi\rangle + \langle d\Psi|d\Psi\rangle = 0$, which is automatically satisfied by the combination $(R - \langle R\rangle)dt$ acting on $|\Psi\rangle$ in Eq. (6.12a). From the evolution equation for $|\Psi\rangle$ given in Eq. (6.12a), we can use the Itô product rule to calculate the evolution equation for the density matrix $\hat{\rho} = |\Psi\rangle\langle\Psi|$, with the result

$$d\hat{\rho} = (d|\Psi\rangle)\langle\Psi| + |\Psi\rangle(d\langle\Psi|) + (d|\Psi\rangle)(d\langle\Psi|)$$

$$= i\hbar^{-1}[\hat{\rho}, H_{\text{eff}}]dt - \frac{1}{2}(\beta_R^2 + \beta_I^2)[H_{\text{eff}}, [H_{\text{eff}}, \hat{\rho}]]dt$$

$$- \frac{1}{2}\gamma\int d^3x([\mathcal{M}^R(\vec{x}), [\mathcal{M}^R(\vec{x}), \hat{\rho}]] + [\mathcal{M}^I(\vec{x}), [\mathcal{M}^I(\vec{x}), \hat{\rho}]])dt$$

$$+ [\hat{\rho}, [\hat{\rho}, Q + R]]dt$$

$$+ \beta_R[\hat{\rho}, [\hat{\rho}, H_{\text{eff}}]]dW_t^R + i\beta_I[H_{\text{eff}}, \hat{\rho}]dW_t^I$$

$$+ \int d^3x([\hat{\rho}, [\hat{\rho}, \mathcal{M}^R(\vec{x})]]dW_t^R(\vec{x}) + i[\mathcal{M}^I(\vec{x}), \hat{\rho}]dW_t^I(\vec{x})), \quad (6.13a)$$

where we have used the identity (for any operator $\mathcal{O}$) $\hat{\rho}\mathcal{O}\hat{\rho} = |\Psi\rangle\langle\Psi|\mathcal{O}|\Psi\rangle\langle\Psi| = \hat{\rho}\langle\mathcal{O}\rangle$. Since the stochastic expectation $E[\ ]$ of all the Brownian differentials is zero

$$E[dW_t^{R,I}] = E[dW_t^{R,I}(\vec{x})] = 0, \quad (6.13b)$$

we find from Eq. (6.13a) that the stochastic expectation of the density matrix obeys the ordinary differential equation

$$\frac{dE[\hat{\rho}]}{dt} = i\hbar^{-1}[E[\hat{\rho}], H_{\text{eff}}] - \frac{1}{2}(\beta_R^2 + \beta_I^2)[H_{\text{eff}}, [H_{\text{eff}}, E[\hat{\rho}]]]$$

$$- \frac{1}{2}\gamma\int d^3x([\mathcal{M}^R(\vec{x}), [\mathcal{M}^R(\vec{x}), E[\hat{\rho}]]] + [\mathcal{M}^I(\vec{x}), [\mathcal{M}^I(\vec{x}), E[\hat{\rho}]]])$$

$$+ [\hat{\rho}, [\hat{\rho}, Q + R]]. \quad (6.13c)$$

An important feature of Eq. (6.13c) is that apart from the final term involving $Q + R$, the stochastic expectation $E[\hat{\rho}]$ obeys a linear master equation of the type studied by Lindblad (1976) and Gorini, Kossakowski, and Sudarshan (1976), which is characteristic quite generally of open system dynamics. If the evolution equation for $E[\hat{\rho}]$ were nonlinear, then as shown by Gisin (1989, 1990), Gisin and Rigo (1995), and Polchinski (1991), there would be a possibility of instantaneous (faster than light) signaling. Assuming that the underlying dynamics is such that faster than light signaling is absent in the emergent quantum theory, the implicit quadratic term R is then fixed to be $-Q$, so that the nonlinear term in Eq. (6.13c) is cancelled to zero. Note that although the evolution of $\hat{\rho}$ in Eq. (6.13a) still contains nonlinearities, they appear only in the unpredictably fluctuating Brownian motion terms, and so cannot be used for faster than light signaling. Taking $Q + R = 0$, our final result for the evolution of the normalized state vector $|\Psi\rangle$ is

$$
\begin{aligned}
|d\Psi\rangle = \Big[&-i\hbar^{-1}H_{\text{eff}}dt - \frac{1}{2}\beta_I^2 H_{\text{eff}}^2 dt - \frac{1}{2}\beta_R^2 (H_{\text{eff}} - \langle H_{\text{eff}}\rangle)^2 dt \\
&- \frac{1}{2}\gamma dt \int d^3x \mathcal{M}^I(\vec{x})^2 - \frac{1}{2}\gamma dt \int d^3x (\mathcal{M}^R(\vec{x}) - \langle \mathcal{M}^R(\vec{x})\rangle)^2 \\
&+ i\beta_I dW_t^I H_{\text{eff}} + \beta_R dW_t^R (H_{\text{eff}} - \langle H_{\text{eff}}\rangle) \\
&+ i\int d^3x dW_t^I(\vec{x})\mathcal{M}^I(\vec{x}) + \int d^3x dW_t^R(\vec{x}) \left(\mathcal{M}^R(\vec{x}) - \langle \mathcal{M}^R(\vec{x})\rangle\right) \Big]|\Psi\rangle,
\end{aligned}
$$

$$(6.14a)$$

and for the evolution of the density matrix $\hat{\rho}$ is

$$
\begin{aligned}
d\hat{\rho} = {} & i\hbar^{-1}[\hat{\rho}, H_{\text{eff}}]dt - \frac{1}{2}(\beta_R^2 + \beta_I^2)[H_{\text{eff}}, [H_{\text{eff}}, \hat{\rho}]]dt \\
&- \frac{1}{2}\gamma \int d^3x \big([\mathcal{M}^R(\vec{x}), [\mathcal{M}^R(\vec{x}), \hat{\rho}]] + [\mathcal{M}^I(\vec{x}), [\mathcal{M}^I(\vec{x}), \hat{\rho}]]\big)dt \\
&+ \beta_R[\hat{\rho}, [\hat{\rho}, H_{\text{eff}}]]dW_t^R + i\beta_I[H_{\text{eff}}, \hat{\rho}]dW_t^I \\
&+ \int d^3x \big([\hat{\rho}, [\hat{\rho}, \mathcal{M}^R(\vec{x})]]dW_t^R(\vec{x}) + i[\mathcal{M}^I(\vec{x}), \hat{\rho}]dW_t^I(\vec{x})\big). \quad (6.14b)
\end{aligned}
$$

Through the reasoning leading to Eqs. (6.14a,b) we have established a connection between the quantum dynamics emergent from matrix model dynamics, and a large body of literature dealing with stochastic modifications to the Schrödinger equation. The "continuous spontaneous localization" or CSL approach of Pearle (1989) and Ghirardi, Pearle, and Rimini (1990) corresponds to taking $\mathcal{M}^R(\vec{x})$ to have the structure

$$
\mathcal{M}^R(\vec{x}) = \int d^3y\, g(\vec{x} - \vec{y}) \sum_r m_r N_r(\vec{y}), \tag{6.14c}
$$

with $g(\vec{x})$ a spherically symmetric, sharply peaked correlation function and with $N_r(\vec{y})$ the number density of particle species r. (It is then natural to assume a similar structure for $\mathcal{M}^I(\vec{x})$, but as we shall see this term does not by itself lead to state vector reduction.) The "energy driven" approach of Bedford and Wang (1975, 1977), Milburn (1991), Percival (1995, 1998), Hughston (1996), and Fivel (1997), on the other hand, is obtained by taking $\mathcal{M}^R(\vec{x}) = gH_{\text{eff}}$, with g a proportionality constant (again, with this assumption, it is natural to assume a similar structure for $\mathcal{M}^I(\vec{x})$). There is a very large literature dealing with objective state vector reduction, which we shall not attempt to review here. For recent reviews focusing on the spontaneous localization approach, and related references, see Bassi and Ghirardi (2003) and Pearle (1999b), while for detailed mathematical and phenomenological studies of the case in which the stochasticity is driven by the Hamiltonian see Adler *et al.* (2001), Adler and Horwitz (2000), and Adler (2002, 2003a). The connection between general Gaussian noise, and the simpler case of white noise, has been discussed by Pearle (1993, 1995) and by Bassi and Ghirardi (2002). Looking back from an historical perspective, the seminal ideas in the stochastic reduction program arose from work over the last twenty-five years by Pearle (1976, 1979, 1982, 1984, 1989, 1990), Ghirardi, Rimini, and Weber (1986), Ghirardi, Pearle, and Rimini (1990), Gisin (1984, 1989, 1990), Diósi (1988a,b, 1989), and Percival (1994, 1995, 1998). The main result coming from the stochastic Schrödinger program, as embodied in Eqs. (6.14a,b), is that when the real (superscript R) noise terms are nonzero, and the operator coefficients of the noise terms are mutually commutative and commute with H_{eff}, the state vector reduces on the eigenstate basis that diagonalizes the operator coefficients of the noise terms, with reduction probabilities given by the Born rule or (in the case of degeneracies) the Lüders rule. A detailed proof of this is given in Section 6.2, and a phenomenological discussion of the localization and energy-driven models for objective state vector reduction is given in Sections 6.3 through 6.5, leading to the conclusion that Eqs. (6.14a,b) provide a theoretically and experimentally viable approach to the state vector reduction problem. Thus, the statistical mechanics of matrix models with a global unitary invariance can lead not only to an emergent complex quantum mechanics, but also to the emergence of the usual probabilistic framework needed for the application of quantum theory.

One feature of stochastic Schrödinger equations that has led to much discussion in the literature is the fact that the usual formulations are nonrelativistic, and attempts to construct relativistic generalizations have encountered serious obstacles (see, e.g., Pearle, 1990, 1999a and Adler and Brun, 2001). Within the framework given here, this is not surprising, since the canonical ensemble that we have used to derive emergent quantum mechanics picks out a preferred rest frame, which we have tentatively identified with the rest frame of the cosmological blackbody radiation. In the decoupling limit in which the τ terms are neglected and in which fluctuations in $\tilde{C}$ are neglected, we have argued that a Lorentz invariant effective quantum theory results when the underlying trace Lagrangian is Lorentz invariant. However, in order for fluctuations in the canonical ensemble to have a finite magnitude, the convergence factor $\exp(-\tau \mathbf{H})$ in the canonical ensemble is needed, and so fluctuation processes in the ensemble are necessarily frame dependent. From this point of view, the frame-dependent structure of the stochastically modified Schrödinger equation is a natural feature.

6.2 Proof of reduction with Born rule probabilities

We give here the proof, following ideas in Ghirardi, Pearle, and Rimini (1990), Hughston (1996), and Adler and Horwitz (2000), that with appropriate choices of the operators appearing as coefficients of the Brownian motion terms, Eqs. (6.14a,b) imply state vector reduction with Born rule probabilities. For ease of exposition, we consider a simplified version of Eqs. (6.14a,b) in which a single operator A that commutes with H_{eff} appears as coefficient of the Brownian motion terms, and then will state the generalization of our results to the more realistic multi-operator case at the end of the argument. Thus, we take for the stochastic Schrödinger equations obeyed by $|\Psi\rangle$ and its adjoint

$$
\begin{aligned}
d|\Psi\rangle = {}&- i\hbar^{-1} H_{\text{eff}}|\Psi\rangle dt - \frac{1}{2}[\beta_R^2(A - \langle A\rangle)^2 + \beta_I^2 A^2]|\Psi\rangle dt \\
&+ \beta_R(A - \langle A\rangle)|\Psi\rangle dW_t^R + i\beta_I A|\Psi\rangle dW_t^I,
\end{aligned}
$$

$$
\begin{aligned}
d\langle\Psi| = {}&\langle\Psi|i\hbar^{-1} H_{\text{eff}} dt - \langle\Psi|\frac{1}{2}[\beta_R^2(A - \langle A\rangle)^2 + \beta_I^2 A^2]dt \\
&+ \langle\Psi|\beta_R(A - \langle A\rangle)dW_t^R - \langle\Psi|i\beta_I A dW_t^I,
\end{aligned}
$$

(6.15a)

and as the corresponding density matrix evolution (abbreviating $|\beta|^2 = \beta_R^2 + \beta_I^2$)

$$
d\hat{\rho} = i\hbar^{-1}[\hat{\rho}, H_{\text{eff}}]dt - \frac{1}{2}|\beta|^2[A,[A, \hat{\rho}]]dt + \beta_R[\hat{\rho},[\hat{\rho}, A]dW_t^R + i\beta_I[A, \hat{\rho}]dW_t^I.
$$

(6.15b)

We begin by remarking that for any operator G commuting with H_{eff} and with A, we have

$$
\begin{aligned}
E[d\langle G\rangle] &= E[\text{Tr}Gd\hat{\rho}] = \text{Tr}G E[d\hat{\rho}] \\
&= \text{Tr}G \left(i\hbar^{-1}[E[\hat{\rho}], H_{\text{eff}}] - \frac{1}{2}|\beta|^2[A, [A, E[\hat{\rho}]]] \right) dt \\
&= \text{Tr}\left(-i[G, H_{\text{eff}}]\hbar^{-1}E[\hat{\rho}] - \frac{1}{2}|\beta|^2[G, A][A, E[\hat{\rho}]] \right) dt = 0,
\end{aligned}
$$

(6.16a)

where as before $E[\ ]$ denotes the expectation with respect to the stochastic process, with $E[dW_t^R] = E[dW_t^I] = 0$. Consider now the variance of A (the square of its uncertainty), defined by

$$V = \langle (A - \langle A \rangle)^2 \rangle = \mathrm{Tr}\hat{\rho}A^2 - (\mathrm{Tr}\hat{\rho}A)^2. \tag{6.16b}$$

Using the Itô product rule of Eq. (6.11a), together with Eqs. (6.15b) and (6.16a), we have

$$
\begin{aligned}
dE[V] &= E[dV] \\
&= E[\mathrm{Tr}d\hat{\rho}A^2 - 2\mathrm{Tr}\hat{\rho}A\,\mathrm{Tr}d\hat{\rho}A - (\mathrm{Tr}d\hat{\rho}A)^2] \\
&= - E[(\mathrm{Tr}d\hat{\rho}A)^2] \\
&= - \beta_R^2 E[(\mathrm{Tr}([\hat{\rho}, A])^2)^2]dt = -4\beta_R^2 E[V^2]dt.
\end{aligned}
\tag{6.16c}
$$

In going from the second to the third line of Eq. (6.16c), we have used the fact that $E[FdW_t^R] = 0$ for any stochastic function F defined at time t (the Itô differential dW_t^R refers to the time interval from t to $t + dt$, and so is statistically independent of functions involving the stochastic history only up to time t). Integrating with respect to time, we see that the expectation $E[V]$ satisfies the integral equation

$$E[V(t)] = E[V(0)] - 4\beta_R^2 \int_0^t ds\,E[V(s)^2], \tag{6.16d}$$

which using the inequality $0 \le E[(V - E[V])^2] = E[V^2] - E[V]^2$ gives the inequality

$$E[V(t)] \le E[V(0)] - 4\beta_R^2 \int_0^t ds\,E[V(s)]^2. \tag{6.16e}$$

Since the variance V is necessarily non-negative, Eq. (6.16e) implies that $E[V(\infty)] = 0$, and again using non-negativity of V this implies that $V(s)$ vanishes as $s \to \infty$, apart from a set of outcomes occurring with probability zero. Thus as $t \to \infty$, the stochastic Schrödinger equation of Eq. (6.15a) drives $|\Psi\rangle$ to a definite A eigenstate when the eigenvalues of A are nondegenerate, which for the time being we assume. (We shall consider the degenerate case shortly.) Rewriting the differential form of the inequality of Eq. (6.16e) as

$$-\frac{dE[V(t)]}{dt}E[V(t)]^{-2} = \frac{dE[V(t)]^{-1}}{dt} \ge 4\beta_R^2, \tag{6.16f}$$

integrating with respect to time, and using the fact that $E[V(0)] = V(0)$ since the stochastic effects act only after $t = 0$, we get the inequality (Hughston, 1996)

$$E[V(t)] \le \frac{V(0)}{1 + 4\beta_R^2 V(0)t}. \tag{6.16g}$$

This gives an explicit bound on $E[V(t)]$ as a function of time, and will be used in the next section to estimate reduction rates.

To see that Born rule probabilities emerge, we apply Eq. (6.16a) to the projectors $\Pi_a \equiv |a\rangle\langle a|$ on a complete set of A eigenstates $|a\rangle$. Since we have assumed that A commutes with H_{eff}, these projectors all commute with H_{eff} and with A, and so by Eq. (6.16a) the expectations $E[\langle \Pi_a\rangle]$ are time independent; additionally, by completeness of the states $|a\rangle$, we have $\sum_a \langle \Pi_a\rangle = 1$. But these are just the conditions for Pearle's (1982, 1989) gambler's ruin or martingale argument to apply. At time zero, when the stochastic evolution has just started, $E[\langle \Pi_a\rangle] = \langle \Pi_a\rangle \equiv p_a$ is the absolute value squared of the quantum mechanical amplitude to find the initial state in A eigenstate $|a\rangle$. At $t = \infty$, the system always evolves to a definite A eigenstate, with the eigenstate $|f\rangle$ occurring with some probability P_f. The expectation $E[\langle \Pi_a\rangle]$, evaluated at infinite time, is then

$$E[\langle \Pi_a\rangle] = \sum_f \langle f|\Pi_a|f\rangle P_f$$

$$= 1 \times P_a + \sum_{f\neq a} 0 \times P_f = P_a; \tag{6.17a}$$

hence $p_a = P_a$ for each a and the state collapses into energy eigenstates at $t = \infty$ with probabilities given by the usual quantum mechanical Born rule applied to the initial wave function. It is also instructive to write Eq. (6.17a) as

$$E[\langle \Pi_a\rangle] = \text{Tr}E[\hat{\rho}(\infty)]\Pi_a = P_a, \tag{6.17b}$$

which exhibits the role of

$$E[\hat{\rho}(\infty)] = \sum_f P_f|f\rangle\langle f| \tag{6.17c}$$

as the conventionally defined mixed state density matrix. We see that during the collapse process, the expectation $E[\hat{\rho}(t)]$ of the pure state density matrix $\hat{\rho}(t)$ with respect to the Brownian fluctuations, evolves from a pure state density matrix at $t = 0$ to a mixed state density matrix at later times.

Let us now consider the case in which the operator A is degenerate. In this case, let us choose a basis of A eigenstates so that, within each degenerate manifold, one basis element coincides (after normalization) with the projection of the initial state vector into that manifold, and the others are orthogonal to it. (If the projection of the initial state vector into the manifold vanishes, any orthonormal basis for that manifold suffices.) We can then apply the argument just given for the nondegenerate case, using this specially chosen A eigenstate basis. We learn that the state vector reduces to one of the members of this basis, with a probability equal to the modulus squared of the projection of the initial state vector on this basis. Thus, the state vector reduces into one or another of the degenerate A eigenstate manifolds, with the result of reduction being the normalized projection of the initial state vector into that manifold, and with the probability of obtaining this outcome equal to the squared modulus of the projection of the initial state into the manifold (Gisin, 1984; Adler *et al.*, 2001). This is precisely the result expected from the Lüders projection postulate, which generalizes the Born rule to the degenerate case. Heuristically, the reason the Lüders rule arises from Eq. (6.15a) is that this equation has the form $d|\Psi\rangle = \mathcal{O}|\Psi\rangle$, with $\mathcal{O}$ diagonal on an A basis. Thus any A eigenstate component that has coefficient zero in the eigenstate expansion of the initial state vector cannot obtain a non-zero coefficient through the subsequent stochastic evolution.

Referring to Eqs. (6.16c–e), we see that state vector reduction occurs only when $\beta_R \neq 0$, since the β_I term in Eq. (6.15a) does not contribute to the evolution of the variance of A. (The necessity for having $\beta_R \neq 0$ to achieve reduction motivates the detailed discussion in Section 6.1 of how to achieve non-zero values for the analogous quantities $\mathcal{K}_1$ and $\mathcal{N}_1(t)$.) When $\beta_R = 0$ Eq. (6.17a) remains valid, but the variance of A remains constant, and so in this case the system stays for all time in the same superposition of A eigenstates that was present at time $t = 0$, apart from the insertion of stochastic phases proportional to β_I in each A eigenstate component.

Finally, we apply the proof just given to the energy-driven and localization models for reduction formulated in Section 6.1. In the case of the energy-driven model, the reduction proof is immediately applicable with the choice $A = H_{\text{eff}}$, and shows that the initial state vector reduces to energy eigenstates with Born rule probabilities. To treat the localization model, we note that the proof just given readily generalizes to the case of multiple noise terms, the operator coefficients of which form a mutually commuting set with one another and with H_{eff}. For the localization model formulated in Section 6.1, the number density operators $N_r(\vec{y})$ do form a mutually commuting set, but in general do not commute with the Hamiltonian H_{eff}. Thus the proof just given demonstrates reduction to eigenstates of the mutually commuting set of number density operators, with Born rule probabilities, in the approximation in which the Hamiltonian-driven evolution of the state vector can be neglected during the reduction time. Actually, in two circumstances stronger statements can be made. First, when the Hamiltonian can be well approximated by its rest mass terms, as in the model of Eq. (6.6), it then commutes with the number density operators, and so the reduction proof applies. Second, even without this approximation, it is often the case that the initial state is a superposition of states that have degenerate energies E, in which case the Hamiltonian H_{eff} can be replaced in its action on the initial state by the operator $1_M E$, with 1_M the unit operator on the degenerate energy manifold M spanned by the components of the initial state. In this case, the Hamiltonian and number density operators effectively commute, so that the reduction proof can be used. This second scenario applies, for example, when the initial state is a superposition of apparatus states with identical energies, which are distinguishable by their macroscopic spatial displacement.

6.3 Phenomenology of stochastic reduction – reduction rate formulas

We turn now to a discussion of phenomenological aspects of the stochastic Schrödinger equation of Eq. (6.14a). Since we have seen in the preceding section that the imaginary noise terms do not contribute to state vector reduction, we shall simplify the discussion by omitting these terms. We shall also change notation by writing $\beta_R = \frac{1}{2}\sigma$, since this is the notation employed in much of the literature on the energy-driven stochastic equation, and we shall henceforth set $\hbar$ equal to 1. With these changes, Eq. (6.14a) becomes

$$
|d\Psi\rangle = \Big[-iH_{\text{eff}}dt - \frac{1}{8}\sigma^2(H_{\text{eff}} - \langle H_{\text{eff}}\rangle)^2 dt
$$
$$
- \frac{1}{2}\gamma dt \int d^3x \big(\mathcal{M}^R(\vec{x}) - \langle \mathcal{M}^R(\vec{x})\rangle\big)^2
$$
$$
+ \frac{1}{2}\sigma dW_t^R(H_{\text{eff}} - \langle H_{\text{eff}}\rangle)
$$
$$
+ \int d^3x dW_t^R(\vec{x})\big(\mathcal{M}^R(\vec{x}) - \langle \mathcal{M}^R(\vec{x})\rangle\big)\Big]|\Psi\rangle. \tag{6.18a}
$$

We shall actually consider separately the cases of energy-driven stochastic reduction, for which Eq. (6.18a) takes the form

$$
|d\Psi\rangle = \Big[-iH_{\text{eff}}dt - \frac{1}{8}\sigma^2(H_{\text{eff}} - \langle H_{\text{eff}}\rangle)^2 dt + \frac{1}{2}\sigma dW_t^R(H_{\text{eff}} - \langle H_{\text{eff}}\rangle)\Big]|\Psi\rangle
$$
$$
\tag{6.18b}
$$

and localization-driven reduction, for which Eq. (6.18a) takes the form

$$
|d\Psi\rangle = \Big[-iH_{\text{eff}}dt - \frac{1}{2}\gamma dt \int d^3x (\mathcal{M}^R(\vec{x}) - \langle \mathcal{M}^R(\vec{x})\rangle)^2
$$
$$
\tag{6.18c}
$$
$$
.+ \int d^3x dW_t^R(\vec{x})\big(\mathcal{M}^R(\vec{x}) - \langle \mathcal{M}^R(\vec{x})\rangle\big)\Big]|\Psi\rangle.
$$

From Eq. (6.16g), we get inequalities (Hughston, 1996) that govern the reduction rate in the two cases . For the energy-driven case of Eq. (6.18b), writing

$$
V = \langle (H_{\text{eff}} - \langle H_{\text{eff}}\rangle)^2\rangle, \tag{6.19a}
$$

we get

$$\frac{E[V(t)]}{V(0)} \leq \frac{1}{1 + \sigma^2 V(0)t}, \tag{6.19b}$$

and so the reduction rate is bounded in terms of the initial state energy variance $(\Delta E)^2 = V(0)$ by

$$\Gamma_R \geq \sigma^2 (\Delta E)^2. \tag{6.19c}$$

For the localization case of Eq. (6.18c), writing

$$V = \langle \int d^3x (\mathcal{M}^R(\vec{x}) - \langle \mathcal{M}^R(\vec{x}) \rangle)^2 \rangle, \tag{6.20a}$$

we get

$$\frac{E[V(t)]}{V(0)} \leq \frac{1}{1 + 4\gamma V(0)t}, \tag{6.20b}$$

and so the reduction rate is bounded in terms of the initial state variance $\int d^3x (\Delta \mathcal{M}^R(\vec{x}))^2 = V(0)$ by

$$\Gamma_R \geq 4\gamma \int d^3x (\Delta \mathcal{M}^R(\vec{x}))^2, \tag{6.20c}$$

with the continuous spontaneous localization model corresponding to the choice for $\mathcal{M}^R(\vec{x})$ given in Eq. (6.14c). Using these estimates of the reduction rate, we shall address two principal issues. The first is to determine bounds on the stochastic terms implied by the agreement of current experiments with the predictions of quantum mechanics, including an analysis of experiments in which coherence is observed in multi-particle systems. The second is to see whether, consistent with these bounds, the models actually explain state vector reduction in measurement situations.

6.4 Phenomenology of energy-driven reduction

We begin with a discussion of the energy-driven reduction equation of Eq. (6.18b). Because at present no deviations from the predictions of quantum mechanics have been observed, there will clearly be experimental constraints on the magnitude of the stochasticity parameter σ. These can be analyzed by using the bound on the reduction rate given in Eq. (6.19c). Since we see from this equation (or directly from the stochastic Schrödinger equation) that σ has units $(\text{mass})^{-\frac{1}{2}}$, it will be convenient to write $\sigma = M^{-\frac{1}{2}}$, with M a characteristic mass scale for the fluctuations that give rise to the stochasticity in the Schrödinger equation. Equation (6.19c) then takes the form

$$\Gamma_R \geq (\Delta E)^2 / M, \qquad (6.21a)$$

and we will quote constraints on σ in the form of lower bounds on M.

In addition to Eq. (6.21a), we shall also need the results of an analysis of stochastic effects on the Weisskopf–Wigner analysis of decay rates and line shapes given in Adler (2003a), which incorporates some simplifying observations of Diósi (personal communication, 2002). Diósi observes that expectations of probabilities can be calculated directly from the expectation of the density matrix, which from Eq. (6.14b) (with $\mathcal{M}^{(R,I)} = 0$) has the same form for real ($\beta_R \neq 0$) or imaginary ($\beta_I \neq 0$) stochastic noise terms. Hence expectations of probabilities arising from the real noise equation Eq. (6.18b) will be the same as those arising from the imaginary noise equation

$$d|\Psi\rangle = -i H_{\text{eff}}|\Psi\rangle dt - \frac{1}{8}\sigma^2 H_{\text{eff}}^2|\Psi\rangle dt + \frac{1}{2}i\sigma H_{\text{eff}}|\Psi\rangle dW_t, \qquad (6.21b)$$

with $dW_t = dW_t^I$ normalized as $dW_t^2 = dt$, which it is easy to show can be formally integrated to give

$$|\Psi(t)\rangle = \exp\left[-i H_{\text{eff}}\left(t - \frac{1}{2}\sigma W_t\right)\right]|\Psi(0)\rangle, \qquad (6.21c)$$

where $W_t = \int_0^t dW_u$ is the Brownian motion with differential dW_t. These considerations lead to the following simple rule for calculating the stochastic modifications of the expectations of probabilities governed by Eq. (6.18b). Let $E[P^\sigma(t)]$ be the expectation of a probability of physical interest, viewed as a function of σ as well as of t, so that $E[P^0(t)] = P^0(t)$ is the value calculated from the standard Schrödinger evolution with no stochasticity. Here t is the elapsed time from the formation of a non-stationary state, such as a metastable decaying state, or more generally, any given superposition of different energy eigenstates. Then Eqs. (6.21b,c) imply the simple relation

$$E[P^\sigma(t)] = E\left[P^0\left(t - \frac{1}{2}\sigma W_t\right)\right], \qquad (6.21d)$$

between the probability calculated in the standard Schrödinger analysis, and the stochastic expectation of the corresponding probability as calculated from Eq. (6.18b), both starting from the same given initial state. The recipe is simply this: take the known expression for the probability calculated in standard quantum mechanics, replace t by $t - \frac{1}{2}\sigma W_t$, and take the stochastic expectation. The needed stochastic expectations of powers of W_t are readily calculated from the Itô rules of Eqs. (6.10c), (6.11a), and (6.13b), which can be used to derive the "exponential martingale" formula

$$E[\exp(\sigma W_t)] = \exp\left(\frac{1}{2}\sigma^2 t\right), \tag{6.21e}$$

the Taylor expansion of which gives

$$E[W_t] = 0, \quad E[W_t^2] = t, \quad E[W_t^3] = 0, \quad E[W_t^4] = 3t^2, \ldots \tag{6.21f}$$

(An alternative method is to use the fact that, for any function $f(W_t)$, the expectation over the Brownian noise is given by

$$E[f(W_t)] = (2\pi t)^{-\frac{1}{2}} \int_{-\infty}^{\infty} dW_t \exp\left(-W_t^2/(2t)\right) f(W_t), \tag{6.21g}$$

which gives $E[W_t^{2n}] = t^n (2n-1) \cdot (2n-3) \cdots 1$.)

As an example of the application of Eqs. (6.21d,e), let us consider the short time survival probability of a decaying state $|s(t)\rangle$. When $\sigma = 0$ we have for the standard answer, which gives the quantum Zeno effect (Misra and Sudarshan, 1977; for recent discussions see Facchi and Pascazio, 2003 and Giulini *et al.*, 1996)

$$|\langle s(0)|s(t)\rangle|^2 = 1 - \langle (H - \langle H \rangle_s)^2 \rangle_s t^2 + O(t^3), \tag{6.22a}$$

with $\langle \mathcal{O} \rangle_s \equiv \langle s(0)|\mathcal{O}|s(0)\rangle$. Following the recipe, we obtain

$$E\left[\left(t - \frac{1}{2}\sigma W_t\right)^2\right] = E\left[t^2 - t\sigma W_t + \frac{1}{4}\sigma^2 W_t^2\right] = t^2 + \frac{1}{4}\sigma^2 t. \tag{6.22b}$$

On substitution into Eq. (6.22a) this gives for general σ

$$E[|\langle s(0)|s(t)\rangle|^2] = 1 - \langle (H - \langle H \rangle_s)^2 \rangle_s \left(t^2 + \frac{1}{4}\sigma^2 t\right) + \ldots \tag{6.22c}$$

We see that there is a stochastic suppression of the quantum Zeno effect, since the short time survival probability never has a regime in which there is a t^2 behavior. The stochastically induced t term in Eq. (6.22c) has precisely the form one would expect from the stochastically induced reduction rate Γ_R estimated in Eq. (6.19c).

Applying the same reasoning to the Weisskopf–Wigner analysis (Adler, 2003a) of decay processes, one learns that within the limits of their approximations, the transition rate per unit time Γ, and the standard Lorentzian line profile formula, are unaffected by the stochastic terms in Eq. (6.18b). Similarly, applying the rule of Eq. (6.21d) to the Rabi oscillations of a two-level system (see Rabi, 1937; Rabi, Ramsey, and Schwinger, 1954; and Feynman, Vernon, and Hellwarth, 1957), that is started precessing at $t = 0$ in a constant magnetic field with Rabi frequency Ω and an initial density matrix $\rho = (1 - \vec{R} \cdot \vec{\tau})/2$ (where $\vec{\tau}$ are the standard Pauli matrices), one finds that

$$E[\vec{R}^{\sigma}(t)] = \exp\left(-\frac{1}{8}\Omega^2\sigma^2 t\right)\vec{R}^0(t). \tag{6.23a}$$

This implies that the expected probabilities for finding the system in the upper (+) and lower (−) state, when the system evolves under the stochastic Schrödinger equation of Eq. (6.18b), are related to the corresponding $\sigma = 0$ probabilities by

$$\frac{1}{2} - E[P_\pm^\sigma] = \exp\left(-\frac{1}{8}\Omega^2\sigma^2 t\right)\left(\frac{1}{2} - P_\pm^0\right). \tag{6.23b}$$

Let us now apply the formulas of Eq. (6.21a) through Eq. (6.23b) to estimate bounds on σ implied by the good agreement of standard quantum theory with experiment. We first discuss the constraints imposed by the maintenance of coherence.

According to Eq. (6.21a), the sole criterion governing how rapidly the state vector reduces is the energy variance; whether the system is microscopic or macroscopic plays no role. We can rephrase Eq. (6.21a) as giving a lower bound on the mass M in terms of the time t_C over which a superposition of energy states differing by ΔE is observed to remain coherent

$$M > t_C(\Delta E)^2, \tag{6.24a}$$

where we have assumed that the stochastic evolution starts at the beginning of the time interval over which coherence is observed. For example, consider the recent superconducting quantum interference device (SQUID) experiments (Friedman *et al.*, 2000; van der Wal *et al.*, 2000) that observe the existence of coherent superpositions of macroscopic states consisting of oppositely circulating supercurrents. The variance ΔE in the Friedman *et al.* experiment is roughly 8.6×10^{-6}eV, with the circulating currents each corresponding to the collective motion of $\sim 10^9$ Cooper pairs. Using the largest coherence times $t_C \sim 5 \times 10^{-6}$ s observed to date (Vion *et al.*, 2002; Yu *et al.*, 2002), and taking the stochastic evolution to start at the beginning of the observed coherence interval, we get a lower bound of $M > 3.5 \times 10^{-16}(\text{eV})^2\text{s} \sim 5 \times 10^{-10}$GeV. If experiments attain coherence times of a millisecond, which may ultimately be feasible (J. Lukens, personal communication, 2002), this bound would be improved to $M > 10^{-7}$GeV.

In the particle physics realm the most straightforward cases to consider are those involving oscillations of neutrinos, K-mesons, or B-mesons, since these can be treated as two-state systems with negligible interaction with the electromagnetic field, and so Eq. (6.24a) can be directly applied. Again, we assume that the stochastic evolution starts with the creation of the oscillating system that is a coherent superposition of energy eigenstates. For a two-state system with mass splitting

Δm, and mean energy E for the components, one has $\Delta E = \Delta m^2/(2E)$, and so we get from Eq. (6.24a) the estimate

$$M > \frac{t_C (\Delta m^2)^2}{4E^2} = t_C (\Delta m)^2 \left(\frac{m}{E}\right)^2. \tag{6.24b}$$

For neutrinos (Lisi, Marrone, and Montanino, 2000; Adler, 2000), taking the coherence time t_C to be the oscillation time $2\pi/(\Delta E) = 4\pi E/(\Delta m^2)$, Eq. (6.24b) becomes

$$M > 2\pi \Delta E = \frac{\pi \Delta m^2}{E}, \tag{6.24c}$$

which for the parameters appropriate to the atmospheric ($\Delta m^2 \sim 3 \times 10^{-3} (\text{eV})^2$, $E \sim 1 \text{GeV}$) and solar ($\Delta m^2 \sim 7 \times 10^{-5} (\text{eV})^2$, $E \sim 1\text{MeV}$) neutrino oscillation observations, gives the respective estimates of $M > 10^{-20} \text{GeV}$ and $M > 2 \times 10^{-19} \text{GeV}$. For K- and B-mesons at rest in the lab frame, taking the coherence time t_C to be the lifetime τ_S of the shorter-lived component (which is similar in magnitude to the oscillation time), Eq. (6.24a) becomes

$$M > \tau_S (\Delta m)^2, \tag{6.24d}$$

which for the parameters appropriate to the K-meson system ($\Delta m \sim 4 \times 10^{-6} \text{eV}$, $\tau_S \sim .9 \times 10^{-10}$ s) and to the B-meson system ($\Delta m \sim 3 \times 10^{-4} \text{eV}$, $\tau_S \sim 1.6 \times 10^{-12}$ s) gives respective bounds of $M > 2 \times 10^{-15}$ GeV and $M > 2 \times 10^{-13}$ GeV. The K- and B-meson systems give better bounds than are obtained from neutrinos because for the mesons m/E is of order unity, whereas for neutrinos m/E is very small.

Another estimate can be obtained by applying the Rabi oscillation formula of Eq. (6.23b) to the experiment of Itano *et al.* (1990), who carry out a proposal of Cook (1988) to make repeated measurements of a two-level system while the vector $\vec{R}$ is precessing for a time interval $t = \pi/\Omega$, for which the exponential damping factor in Eq. (6.23b) becomes $\exp(-\frac{1}{8}\pi\Omega\sigma^2)$. Corresponding to the experimental value $\Omega = 320.7\text{MHz}$ and the fact that probabilities were observed to an accuracy of about .02 in this experiment, and were found to agree with standard Schrödinger theory, we get a bound of $M > 2 \times 10^{-15} \text{GeV}$, comparable to that obtained from oscillations in the K-meson system.

Let us consider finally bounds coming from particle decay experiments (Adler, 2003a). Because the transition rate per unit time and Lorentzian line shape are unaffected by σ, bounds on σ from particle decays result only from experiments in which a metastable system is monitored as function of time from a known time (or vertex location) of formation. According to the discussion leading to Eq. (6.22c), for small times the effective transition rate per unit time is

$$\Gamma_R = \frac{1}{4}\sigma^2(\Delta E)^2,\qquad(6.25a)$$

with ΔE the initial state energy variance. As already remarked, this can be interpreted as an early time decay rate coming from spontaneous reduction induced by the stochastic fluctuation terms. In order for the rate of Eq. (6.25a) not to lead to pronounced early time deviations from the observed decay rate Γ, we must have

$$\Gamma_R < \Gamma.\qquad(6.25b)$$

Using the standard Golden Rule formula for Γ, and the formula $(\Delta E)^2 = (V^2)_{ss} \equiv \langle s|V^2|s\rangle$ relating the energy variance to the initial state expectation of the square of the decay-inducing perturbation V (assuming $V_{ss} \equiv \langle s|V|s\rangle = 0$, as is generally the case), we obtain the bound

$$M > \frac{(V^2)_{ss}}{4\Gamma} = \frac{\sum_{m\neq s}|V_{sm}|^2}{8\pi \sum_{m\neq s}|V_{sm}|^2\delta(E_m - E_s)} \equiv \frac{E_D}{8\pi},\qquad(6.26)$$

with E_D defining an energy characteristic of the decay process. In a particle physics context, a first guess would be to estimate E_D as being of order the mass of the decaying particle. The most massive decays, for which Γ has been measured by tracking a metastable system from the point of formation, appear to be $\pi^0 \rightarrow \gamma\gamma$ decay, with an initial mass of 140 MeV, and charmed meson decays, with an initial mass of around 2 GeV. Estimating E_D in Eq. (6.26) as the decaying particle mass, these give respective bounds on M of order 6 MeV and 80 MeV, respectively. If M were significantly larger than these bounds, one would have observed anomalous accumulations of decay events close to the production vertex, as a result of decays induced by spontaneous reduction. These bounds are significantly better than those obtained above from experiments that observe the maintenance of coherence in either an atomic or a particle physics context, but are still very weak when compared to the Planck mass of 10^{19}GeV, which as the probable scale for the unification of particle forces with quantum gravity, is a natural scale at which new physics effects are expected.

We note finally that in the analysis of future experiments to improve the phenomenological bounds on M, in addition to the analytical methods discussed here, one may also have recourse to numerical simulation methods (Shack, Brun, and Percival, 1995; Shack and Brun, 1997; Kloeden and Platen, 1997; Percival, 1998). Powerful new techniques for performing such simulations, along with related analytical methods, are given by Brody and Hughston (2002).

We turn now to the second requirement that must be satisfied by a phenomenology of state vector reduction, which is that it should lead to rapid reduction in experimental situations where a probabilistic outcome is observed. According to the von Neumann (1932) model for measurement, a measurement sets up a correlation between states $|f_\ell\rangle$ of a quantum system being measured, and macroscopically distinguishable states $|\mathcal{M}_\ell\rangle$ of the measuring apparatus $\mathcal{M}$, in such a way that an initial state

$$|f\rangle|\mathcal{M}_{\text{initial}}\rangle = \sum_\ell c_\ell |f_\ell\rangle|\mathcal{M}_{\text{initial}}\rangle \qquad (6.27\text{a})$$

evolves unitarily to

$$\sum_\ell c_\ell |f_\ell\rangle|\mathcal{M}_\ell\rangle. \qquad (6.27\text{b})$$

An objective state vector reduction model must then account for the selection of *one* of the alternatives $|f_\ell\rangle|\mathcal{M}_\ell\rangle$ from this superposition, with a probability given by $|c_\ell|^2$. From Eq. (6.21a), we see that in order for reduction to occur within a time t_R, given the mass $M = \sigma^{-2}$ determining the stochasticity parameter, the energy spread ΔE between different states in the superposition of Eq. (6.27b) must be at least

$$\Delta E = (M/t_R)^{\frac{1}{2}}. \qquad (6.27\text{c})$$

Thus, assuming that the relevant experimental measurement time is of order 10^{-8} s, which sets an upper bound on t_R, and using the bound $M > .1\text{GeV} \sim 10^{-7}(\text{eV})^2$ s that we obtained above, we find that ΔE must be greater than a few eV for reduction to occur within the measurement time. (Of course, for values of M larger than this lower bound, ΔE must be correspondingly larger by a factor of $(M/.1\text{GeV})^{\frac{1}{2}}$. For example, if we take M as the Planck mass, ΔE would have to be around 30GeV to give a 10^{-8} s reduction time.) Since in typical molecular beam experiments the energy spread among the states $|f_\ell\rangle$ can be very small fractions of an eV, these energy spreads by themselves cannot quantitatively account for state vector reduction. The only way for reduction to occur within typical measurement times is for the energy spreads among the alternative apparatus states in the superposition to be much larger than a few eV. Since in the ideal measurement model there is no energy transfer from the microscopic system to the apparatus, such an energy spread in the measurement apparatus states can be present, if at all, only through the effects of environmental interactions. At this point the analysis becomes controversial. Hughston (1996) has suggested that accretion of environmental molecules on the apparatus gives a sufficient energy spread between measurement outcomes to drive reduction, and this has been followed up by a detailed study by Adler (2002). However, Pearle (2003) and others have pointed out that the accretion mechanism involves erecting an artificial boundary defining the apparatus as opposed to the environment, and if the analysis is based on a closed as opposed to an open system, there is no accretion-induced energy spread between different outcomes, and no energy-driven state vector reduction! To illustrate these issues, we first present the argument for environmentally induced energy variances driving reduction, as given in Adler (2002), and then give a discussion of problems with this scenario.

For environmental interactions to be effective in producing state vector reduction, they must lead to energy fluctuations ΔE of the apparatus in the course of a measurement, that are large enough for Eq. (6.21a) to predict a reduction time $t_R = \Gamma_R^{-1}$ that is less than the time it takes to make the measurement. (This criterion assumes that the energy fluctuations in the apparatus can be interpreted as corresponding to energy differences between the different measurement outcomes. When the surface of accretion is sufficiently remote from the parts of the apparatus that register measurement outcomes, this assumption could be questioned.) In order to make quantitative estimates, we must first make a guess as to the actual magnitude of M. Perhaps the most natural guess is that M should be of order the Planck mass, 10^{19} GeV $\sim 10^{13}(\text{eV})^2$ s, which to give a reduction time of order 10^{-8} s requires, from Eq. (6.27c), $\Delta E \sim 3 \times 10^{10}$eV $= 30$GeV, i.e., of order the mass of a nitrogen molecule. However, the best lower bound on M that we obtained is also consistent with M of order the electroweak mass 2.5×10^2GeV $\sim 2.5 \times 10^{-4}(\text{eV})^2$ s, which to give a reduction time of order 10^{-8} s requires only $\Delta E \sim 100$eV. For an intermediate scale M, of order the geometrical mean of the electroweak mass and the Planck mass, reduction within 10^{-8} s would require $\Delta E \sim 2 \times 10^6$eV $= 2$MeV, of order a few electron masses.

The starting point for the analysis of environmental accretion is the observation of Adler and Horwitz (2000), Adler (2002), and Pearle (2003) that the energy-driven equation has good clustering properties. Specifically, if the total Hamiltonian is the sum of the Hamiltonians for two noninteracting subsystems, so that $H = H_1 + H_2$, and if subsystem 1 is at the end point of its reduction process, then the reduction dynamics for subsystem 2 is independent of the variables referring to subsystem 1. More generally, if one traces over the variables of subsystem 1, the reduction dynamics for subsystem 2 takes the standard energy-driven form governed by H_2. Taking off from this observation, the analysis of Adler (2002) consists of two parts. First, a splitting of the total system is made into environment on the one hand, and measured system and measuring apparatus on the other, which are assumed to be weakly coupled. The clustering property in the absence of environmental coupling permits one to develop a weak-coupling Hartree approximation when environmental couplings are included, giving a modified reduction equation for the measured system plus measuring apparatus. This takes the form of Eq. (6.18b), with a Hartree Hamiltonian consisting of the Hamiltonian for the measured system and measuring apparatus, plus the expectation of the interaction Hamiltonian between the apparatus and the environment. Second, in terms of this Hartree picture, a detailed analysis is given of mechanisms that can give apparatus energy fluctuations large enough to drive reduction with a Planckian M. The analysis of Adler (2002) considers three possible sources of energy fluctuations: thermal energy fluctuations, fluctuations in apparatus mass from particle accretion processes, and fluctuations in apparatus mass from amplified fluctuations in the currents that actuate the indicator devices. Thermal energy fluctuations were found to be unable to produce a 30GeV energy fluctuation within the measurement time. On the other hand, both energy fluctuations from mass accretion, and from amplified current fluctuations, are relevant.

As illustrations of accretion-induced fluctuations, assuming room temperature and atmospheric pressure (760 Torr), the time for one molecule to be accreted onto an area of 1cm^2 is 3×10^{-24} s, while at an ultrahigh vacuum of 10^{-13} Torr it is 3×10^{-8} s. Thus, for an apparatus in the atmosphere at standard temperature and pressure, where the bulk of the accreting atoms are nitrogen molecules, the minimum apparatus area required for one molecule to accrete in a reduction time of 10^{-8} s is 3×10^{-16}cm^2, with the corresponding minimum area needed at a pressure of 10^{-13} Torr equal to 3cm^2. Further estimates of this type can be made. For example, even in the sparsely populated environment of intergalactic space, one concludes that in a typical high precision molecular beam experiment, the reduction time induced by particle accretion on a capsule large enough to enclose the apparatus would be smaller, by at least an order of magnitude, than the measurement time.

Estimates can also be made of energy fluctuations arising from the amplified fluctuations in the currents which actuate experimental indicating or recording devices. Of course, if power sources are included, there are no overall current fluctuations, but power supplies are typically large in area and so when included in the system the accretion analysis just given indicates rapid reduction times. In a typical electrically amplified measurement, a final total charge transfer Ne (with e the charge of an electron) actuates an indicator or recording device. Assuming that the fluctuation in the current is the amplified fluctuation in the initially detected signal, for amplification gain G we have $\Delta N \sim G \times (N/G)^{\frac{1}{2}} = (NG)^{\frac{1}{2}}$. Let us take N to correspond to a charge transfer of 1 milliampere (a voltage change of 10 volts at 10 kΩ impedance) over a 10^{-8} second pulse, so that $N \sim 6 \times 10^7$, and assume a gain $G \sim 10^4$, giving $\Delta N \sim 8 \times 10^5$. Multiplying by the electron mass of $.5 \times 10^{-3}$GeV, we find that the corresponding energy fluctuation is $\Delta E \sim 4 \times 10^2$GeV, which leads to state vector reduction in 5×10^{-11} s. Thus, electric current fluctuations play a significant role in state vector reduction when the "apparatus" is defined to exclude power sources.

If instead we assume a value of M at the electroweak scale or at an intermediate scale, a much smaller ΔE is needed, which can be supplied by accretion of a few electrons at most. However, even with M as small as the lower bound of .1GeV, thermal fluctuations arising from immersion in the 3-degree Kelvin microwave background radiation are ineffective in causing state vector reduction, since for an apparatus of 1cm^2 area the typical energy fluctuation associated with the microwave background flux in 10^{-8} s is only $\Delta E \sim 10^{-1}$eV, which is too small.

The overall conclusion of this analysis is that, assuming the validity of the apparatus–environment distinction, then for conditions under which laboratory experiments are performed, as well as for conditions under which space capsule experiments might be performed in the foreseeable future, state vector reduction times as estimated by Eq. (6.21a) using the particle accretion mechanism are well within experimental measurement times. However, as noted above, this conclusion is controversial, and is subject to the following objections. First of all, it seems somewhat counterintuitive that particle fluxes should have to be invoked to achieve definite measurement outcomes; one would like to think that a Stern–Gerlach apparatus in intergalactic space, contained within layers of distant shielding that totally absorb incident particles, would still function as on earth. This of course raises the issue of how one defines the apparatus and how one defines the environment in the accretion model. If one takes the Stern–Gerlach apparatus and its capsule as the apparatus, but does not include an outer and distant layer of shielding, one concludes that there is no energy-driven reduction; if one has to take the particle absorbing outer layer of shielding into account, then the situation becomes very much like the von Neumann recursion (an apparatus measuring an apparatus ad infinitum), which the objective reduction program is designed to avoid. Moreover, as stressed by Pearle (2003), if one includes the outer layer of shielding in the apparatus definition, then why not also include those particles that will hit the shielding within the measurement time, in which case there is no accretion effect and no energy-driven reduction. In other words, if one demands that there be reduction independent of the apparatus definition, then the accretion model for energy-driven reduction is in jeopardy. As of this writing, we find the objections to the accretion model convincing, and conclude that although the H_{eff}-driven stochastic terms in Eq. (6.14a) are likely to be present, so that the bounds on them developed above are relevant, they are unlikely to be able to account for state vector reduction through the accretion mechanism.

Another objection to energy-driven reduction, discussed by Pearle (2003), is that if there is reduction of isolated systems to energy eigenstates, then the energy spectrum of the superpositions needed for there to be nontrivial dynamics will be narrowed with the passage of time. This leads to a broadening of the time evolution of physical systems, with possible experimental tests, and to the possibility of improved bounds on the mass $M = \sigma^{-2}$ governing the stochasticity.

One other possibility that should be kept in mind is that spacetime geometry fluctuations may play a role in energy-driven reduction. In recent work summarized in Ng (2003), holographic ideas are invoked to derive intrinsic limits, arising from spacetime geometry fluctuations, on the accuracy with which distances, time intervals, and energies can be determined. For energy, Ng proposes the formula $\delta E \sim E(E/10^{19}\text{GeV})^{\frac{2}{3}}$. This formula is meant to apply in the microscopic realm (it clearly makes no sense for energies near or larger than the Planck mass), so let us suppose that it can be applied to a light nucleus from which an apparatus is constructed, with atomic weight of 10GeV, giving an energy uncertainty of 10^{-11}GeV. If we suppose that for large objects these energy fluctuations can be added in quadrature, then for an apparatus containing 10^{24} nuclei we get an energy uncertainty of $10^{12} \times 10^{-11}$GeV $= 10$GeV, which is the right magnitude to give state vector reduction in about 10^{-8} s, for M of order the Planck mass. In making this estimate we are implicitly assuming that the time scale for the 10GeV fluctuations is less than the measurement time of 10^{-8} s. Since this mechanism does not involve a separation between measured system and environment, it may avoid the objections discussed above. However, it must be regarded as very speculative until there is a more detailed understanding of spacetime fluctuations in energy, and particularly of the time scales on which they occur. For a discussion and references relating to other ideas about a possible role for quantum gravity effects in state vector reduction, see Penrose (1996) and Moroz, Penrose, and Tod (1998).

6.5 Phenomenology of reduction by continuous spontaneous localization

We have seen in Section 6.1 that the fluctuation $\Delta\tilde{C}$ in general contains a matrix part $\mathcal{N}$, which in the rest-mass dominated approximation to the Hamiltonian of Eq. (6.6) gives rise to the mass-weighted matrix $\mathcal{M}$ of Eq. (6.8e). The continuous spontaneous localization (CSL) model (in its mass-proportional coupling form) then follows by assuming a spatially correlated noise structure, as in Eq. (6.10d), coupled to the particle number density operators as in Eq. (6.14c). The sharply peaked correlation function $g(\vec{x})$ is assumed to have characteristic radius r_C and to be normalized to have spatial integral unity

$$\int d^3x g(\vec{x}) = 1. \tag{6.28a}$$

Although the CSL literature often assumes a Gaussian form for the correlation function, so that $g(\vec{x}) \propto \exp(-\vec{x}^2/r_C^2)$, Weber (1990) has shown that no particular choice of functional form is needed, and so it seems just as natural to take the correlation function to be an exponential $g(\vec{x}) \propto \exp(-|\vec{x}|/r_C)$. We shall see explicitly, in the estimates given below, that the results are independent of the detailed form of $g(\vec{x})$, and, indeed, are independent of the value of the correlation length r_C, provided that the latter lies between microscopic and macroscopic dimensions. The value $r_C \sim 10^{-5}$cm is typically assumed in the CSL literature. Ultimately, one would want the underlying theory to predict the value of the correlation length, but at the present stage of development this is not possible; hence a challenge for implementing the CSL mechanism is providing a fundamental rationale for a correlation length of the required magnitude.

To study the phenomenology of the localization model, let us make the simplifying assumption that only one species of particles is present, so that we can drop the subscript r in Eq. (6.14c); this is in fact realistic since with mass proportional couplings the contribution from the nucleons in an apparatus dominates that from the electrons. We now follow the discussion given by Ghirardi, Pearle, and Rimini (1990). Since the number density operator $N(\vec{y}) = \Psi_{\text{eff}}^{\dagger}(\vec{y})\Psi_{\text{eff}}(\vec{y})$ obeys the commutation relation $[N(\vec{y}), \Psi_{\text{eff}}^{\dagger}(\vec{x})] = \delta^3(\vec{y} - \vec{x})\Psi_{\text{eff}}^{\dagger}(\vec{y})$, a state

$$|\vec{r}_1, \ldots, \vec{r}_n\rangle \propto \prod_{\ell=1}^{n} \Psi_{\text{eff}}^{\dagger}(\vec{r}_\ell)|\text{vac}\rangle, \tag{6.28b}$$

is an eigenstate of $\mathcal{M}^R(\vec{x})$ with eigenvalue

$$n(\vec{x}) = m \sum_{\ell=1}^{n} g(\vec{x} - \vec{r}_\ell). \tag{6.28c}$$

We now consider the reduction of a state $|\Psi\rangle$ that is a superposition of a state containing a group of n particles at locations $\vec{r}_\ell$, and a state containing a second group of n particles at locations $\vec{r}_\ell'$, with the separations between the two groups larger than the correlation length r_C, so that

$$|\Psi\rangle = \frac{1}{\sqrt{2}}(|\vec{r}_1, \ldots, \vec{r}_n\rangle + |\vec{r}_1', \ldots, \vec{r}_n'\rangle). \tag{6.29a}$$

Then from Eq. (6.28c), we find that the initial state variance $V(0)$ needed in Eqs. (6.20a–c) is

$$V(0)/m^2 = \left\langle \left(\int d^3x \left(\mathcal{M}^R(\vec{x}) - \langle \mathcal{M}^R(\vec{x}) \rangle \right)^2 \right\rangle \right.$$

$$= \int d^3x \left(\frac{1}{2} \left[\left(\sum_{\ell=1}^{n} g(\vec{x} - \vec{r}_\ell) \right)^2 + \left(\sum_{\ell=1}^{n} g(\vec{x} - \vec{r}_\ell') \right)^2 \right] \right.$$

$$\left. - \frac{1}{4} \left[\sum_{\ell=1}^{n} g(\vec{x} - \vec{r}_\ell) + \sum_{\ell=1}^{n} g(\vec{x} - \vec{r}_\ell') \right]^2 \right)$$

$$= \int d^3x \frac{1}{4} \left[\sum_{\ell=1}^{n} g(\vec{x} - \vec{r}_\ell) - \sum_{\ell=1}^{n} g(\vec{x} - \vec{r}_\ell') \right]^2$$

$$\simeq \int d^3x \frac{1}{4} \left[\left(\sum_{\ell=1}^{n} g(\vec{x} - \vec{r}_\ell) \right)^2 + \left(\sum_{\ell=1}^{n} g(\vec{x} - \vec{r}_\ell') \right)^2 \right]$$

$$\simeq \int d^3x \frac{1}{2} \left(\sum_{\ell=1}^{n} g(\vec{x} - \vec{r}_\ell) \right)^2, \tag{6.29b}$$

where in the final two lines we have used the facts that the two groups of particles are separated by more than the correlation length and that the integrals over the sums corresponding to each group are approximately the same. To evaluate the integral in the final line of Eq. (6.29b), we consider two limiting cases. In the first case, corresponding to macroscopic objects, we assume that each group of particles has a spatial extent much larger than the correlation length, but that the interparticle separation is much smaller than the correlation length and that the particles are distributed with average density D_0. Then we have

$$\int d^3x \frac{1}{2} \left(\sum_{\ell=1}^{n} g(\vec{x} - \vec{r}_\ell) \right)^2$$

$$\simeq \int d^3x \frac{1}{2} ng(\vec{x} - \vec{r}_1) \sum_{\ell=1}^{n} g(\vec{r}_1 - \vec{r}_\ell) \tag{6.29c}$$

$$\simeq \int d^3x \frac{1}{2} ng(\vec{x} - \vec{r}_1) D_0 \int d^3y \, g(\vec{r}_1 - \vec{y}) = \frac{1}{2} n D_0.$$

Thus the initial state variance is given by $V(0) \simeq \frac{1}{2} m^2 n D_0$, and the reduction rate is

$$\Gamma_R \geq 2\gamma m^2 n D_0. \tag{6.29d}$$

In the second case, corresponding to microscopic objects such as large molecules, in which each group of particles is tightly bunched within a radius much smaller than the correlation length, we have

$$\int d^3x \frac{1}{2} \left(\sum_{\ell=1}^{n} g(\vec{x} - \vec{r}_\ell) \right)^2$$

$$\simeq \frac{1}{2} n^2 \int d^3x \, g(\vec{x})^2 \tag{6.29e}$$

$$= c'n^2 r_C^{-3},$$

with c' a dimensionless constant of order unity. So in this case the initial state variance is $V(0) = c'm^2n^2r_C^{-3}$, and (writing $c = 4c'$) the reduction rate is

$$\Gamma_R \geq c\gamma m^2 n^2 r_C^{-3}. \tag{6.29f}$$

To bound the stochasticity parameter γ, instead of using experiments that observe coherent superpositions of energy eigenstates as in the discussion of Section 6.4, we must now rely on experiments that observe coherent superpositions of spatially displaced groups of particles. A useful bound comes from experiments (Arndt *et al.*, 1999; Nairz, Arndt, and Zeilinger, 2000; Nairz *et al.*, 2001) that achieve diffraction of C_{60} and C_{70} fullerenes containing $n \sim 10^3$ nucleons on gratings with spacings of 1 to 2.5 times 10^{-5} cm. Corresponding to the beam transit time in these experiments of order 10^{-2} s, for localization not to spoil the diffraction pattern we require $\Gamma_R << 10^2 \, \text{s}^{-1}$, which from the bunched case formula of Eq. (6.29f) gives

$$\gamma << 10^{-4} r_C^3 \, \text{s}^{-1} \text{GeV}^{-2}. \tag{6.30a}$$

Ghirardi, Pearle, and Rimini (1990) assume a correlation length $r_C \sim 10^{-5}$cm, and propose the value

$$\gamma \sim 10^{-30} \text{cm}^3 \, \text{s}^{-1} \text{GeV}^{-2}, \tag{6.30b}$$

well within the upper limit of 10^{-19}cm^3 s^{-1}GeV^{-2} given by Eq. (6.30a) (and also well within other bounds derived in the CSL literature, obtained from the fact that the CSL reduction mechanism leads to small system energy increases induced by coupling to the noise). For a small macroscopic object with $n \sim 10^{13}$ and $D_0 \sim 10^{24}$cm^{-3}, the formula of Eq. (6.29d) then gives a reduction time $\Gamma_R \sim 10^7 \, \text{s}^{-1}$. So any instrument pointer displacement involving at least 10^{13} nucleons gives a reduction time less than typical experimental measurement times.

What we have given here is only a quick sketch of the CSL reduction mechanism, in the context of our analysis of Sections 6.1 through 6.3. For a comprehensive discussion, the reader is referred to the review of Bassi and Ghirardi

(2003), where all aspects of the CSL approach, including objections that have been raised and proposed answers to these objections, are analyzed in detail. Since our introduction of the particle density in Eq. (6.14c) is an assumption, and other assumptions have been made along the way in Section 6.1, we cannot claim to have derived the CSL mechanism. But our arguments at least show that there is a plausible route leading from the underlying trace dynamics to CSL reduction with mass proportional couplings (which as argued by Pearle and Squires, 1994; Collett *et al.*, 1995; and Pearle *et al.*, 1999; is the phenomenologically favored form of the CSL model).

7

Discussion and outlook

In the preceding chapters we have developed a new approach to quantum mechanics, based on the idea that quantum theory is an emergent phenomenon arising from the statistical dynamics of an underlying matrix model. As we have seen, a very rich structure arises from a few simple assumptions about the structure of the underlying trace dynamics, such as its global unitary invariance and the use in thermodynamic calculations of the maximally symmetric canonical ensemble that gives a non-zero value to $\langle \tilde{C} \rangle_{AV}$.

Our proposal suggests answers to a number of the motivating questions raised in the Introduction. Canonical quantization is seen to emerge as a consequence of the equipartition of $\tilde{C}$, rather than being imposed as a postulate. A potential resolution of the measurement problem appears as a consequence of the fact that the underlying dynamics is not unitary, with the unitary dynamics of quantum field theory emerging only as a thermodynamic approximation; when fluctuation corrections to this approximation are taken into account, our arguments suggest an underlying dynamical origin for stochastic Schrödinger equation models for state vector reduction. The underlying dynamics is nonlocal, with the locality of quantum field theory again the result of the thermodynamic approximation; this suggests an amelioration, in the underlying dynamics, of the infinities of quantum field theory, provides a basis for understanding the nonlocal "paradoxes" of quantum theory, and may have implications as well as for early universe cosmology, where it may play a role in establishing the large-scale uniformity of the universe. Our approach may also have connections to recent work on string theory, where matrix models (as reviewed in Taylor, 2001) play a prominent role.

A significant feature of trace dynamics is that scale invariance is manifested through the vanishing of the Lorentz trace of the trace energy-momentum tensor $\mathcal{T}_{\mu}^{\ \mu}$, providing a single number condition. Similarly, we have seen that supersymmetry in trace dynamics implies only the conservation of a trace supercurrent $\mathbf{J}^{\mu}$ (and its conjugate $\bar{\mathbf{J}}^{\mu}$), again providing a single number condition. Neither the vanishing of $T_\mu{}^\mu$ nor the conservation of $\mathbf{J}^\mu$ imply corresponding operator constraints. Thus in a suitably constructed unified trace dynamics theory of the forces, it is possible that either scale invariance or supersymmetry could provide a single number constraint forcing the vanishing of the cosmological constant, without simultaneously forcing the emergent quantum field theory to be either massless or exactly supersymmetric. For a first attempt in this direction, see Adler (1997c), but a detailed working out of the idea, taking into account the necessity for global unitary fixing in the canonical ensemble (as discussed in Section 4.5), has yet to be given.

A large number of other questions are left unanswered. What we have done in the preceding chapters is to establish a general framework, and to formulate specific assumptions and approximations, that lead to the emergence of quantum mechanics as an effective dynamics for low frequency or "slow" degrees of freedom. These observable degrees of freedom are assumed to be separated by a large hierarchy of scale and effectively decoupled, from very high frequency or "fast" degrees of freedom characterizing the underlying dynamics. We know empirically that such a hierarchy of scales exists, but the mechanism that produces it is not known, and we shed no new light on this question. We also note that our framework leads to not one but two copies of quantum field theory, corresponding to the eigenvalues $\pm i$ of i_{eff}; we have not attempted to assign a physical role to the second copy, nor to the additional "off-diagonal" degrees of freedom corresponding to the parts of the underlying matrices that anticommute with i_{eff}. Since the discussions of Section 2.3 and 2.4 and Chapter 3 show that no decomposition with respect to i_{eff} is involved in the coupling of matrix degrees of freedom to a c-number spacetime metric, the second copy and the off-diagonal degrees of freedom both couple to the spacetime metric in a manner similar to the first copy. Hence they act as source terms in the Einstein field equations, and thus may be relevant for the questions of "dark matter" or "dark energy" that are central to current cosmology (but which may, of course, have more mundane explanations). We remark in this connection that in real Hilbert space, discussed briefly in Appendix A, the second copy of quantum field theory is absent. The reason is that in real Hilbert space i_{eff} cannot be diagonalized, and so itself plays the role of the imaginary unit of the emergent complex quantum theory.

We stress that we have not identified a candidate for *the* specific matrix model that realizes our assumptions: this is a task for the future. There may be a number of theories that realize our assumptions, there may be none, in which case we have erected an empty framework, or there may be only one, which could then provide the underlying unified theory of physical phenomena that is the goal of current researches in high-energy physics and cosmology. We have also not committed as to whether the underlying Hilbert space is finite or infinite, or as to whether the number of fundamental matrix degrees of freedom is finite (as suggested by recent ideas, reviewed by Bousso, 2002, on "holographic" interpretations of quantum theory) or is infinite. It is possible that the underlying dynamics may be discrete, and this could naturally be implemented within our framework of basing an underlying dynamics on trace class matrices.

To conclude, we believe that the kinematical framework of trace dynamics, as developed in this book, provides a fruitful new direction for exploration in the search for a unifying theory of particles and forces. Specifically, if quantum theory is an emergent phenomenon, then the sought for unification of gravitation with the quantum field theoretic standard model may not involve "quantizing" gravitation. Rather, the ideas developed in this book suggest, one should seek a common origin for both gravitation and quantum field theory at the deeper level of physical phenomena from which quantum field theory emerges.

Appendices

To keep the discussion of this book self-contained, a number of topics that are briefly mentioned in the text are treated in more detail in the appendices that follow.

The notation of the appendices follows that generally used in the text. Throughout this book, we indicate sums explicitly, *except* that the usual Einstein summation convention is used for sums over Greek letter four-vector and tensor indices, and a summation convention is used for the indices i, j in Section 3.3. Our Minkowski metric convention is $\eta_{\mu\nu} = \mathrm{diag}(1, 1, 1, -1)$, and we have taken the velocity of light to be unity, so that c does not appear in the equations. However, Planck's constant $\hbar$ is generally retained (except in the formulas of Sections 1.5 and 6.3 through 6.5 where we set $\hbar = 1$), because our approach implies that it has a dynamical origin.

Fermionic quantities are represented throughout by Grassmann numbers. Real Grassmann numbers are constructed as products of a basis χ_r of real Grassmann elements, that obey the anticommutative algebra $\{\chi_r, \chi_s\} = 0$ for all r, s, which implies that the square of any Grassmann element vanishes. Complex Grassmann numbers have real and imaginary parts that are real Grassmann numbers. A product of an even number of Grassmann basis elements, multiplied by a complex number coefficient, is called even grade or bosonic, and a product of an odd number of Grassmann basis elements, with a complex number coefficient, is called odd grade or fermionic. Grassmann integration is defined by $\int d\chi_r \chi_s = -\int \chi_s d\chi_r = \delta_{rs}$, and so is effectively the same as differentiation from the left. The seminal reference on applications of Grassmann numbers in physics is Berezin (1966), and a good pedagogical account can be found in Section 1.3 of Cheng and Li (1988). Often in the literature an adjointness convention $(\chi_r \chi_s)^\dagger = \chi_s^\dagger \chi_r^\dagger$ is adopted for all Grassmann matrices irrespective of grade, which implicitly treats the elements $\chi_1, \chi_2, \ldots$ of the Grassmann algebra as operators with the nontrivial transposition rule $(\chi_r \chi_s)^T = \chi_s^T \chi_r^T$. As noted in Section 1.1, we follow the alternate convention in which $(\chi_r \chi_s)^\dagger = (-1)^{g_r g_s} \chi_s^\dagger \chi_r^\dagger$, so that in our convention the Grassmann elements (i.e., 1×1 Grassmann matrices) obey $(\chi_r \chi_s)^T = \chi_r \chi_s$ and have no implicit operator structure.

Appendix A: Modifications in real and quaternionic Hilbert space

In a complex Hilbert space the scalars that are used to form superpositions of Hilbert space vectors are complex numbers. In real Hilbert space the scalars take only real number values, while in quaternionic Hilbert space the scalars can be quaternions of the form $r_0 + r_1 i + r_2 j + r_3 k$, with $r_{0,1,2,3}$ real, and with i, j, k the quaternion imaginary units obeying the non-commutative algebra $i^2 = j^2 = k^2 = -1$ and $ij = -ji = k$, $jk = -kj = i$, $ki = -ik = j$. The distinguishing feature of complex Hilbert space is that there is an anti-self-adjoint c-number $i1$ that commutes with all operators on Hilbert space, and the trace can take complex values. By contrast, in real and in quaternionic Hilbert space, there exists no anti-self-adjoint c-number, since the only operators that commute with all operators on Hilbert space are of the form $r1$, with r real, and hence are self-adjoint. Also, in real and in quaternionic Hilbert space, the trace is real. In real Hilbert space the reality of the trace is self-evident. In quaternionic Hilbert space it follows from the fact that the diagonal sum $\sum_n (B_1 B_2)_{nn}$ does not obey the cyclic property of Eq. (1.1a) as a result of the non-commutativity of the matrix elements $(B_1)_{nm}$ and $(B_2)_{mn}$ in the quaternion algebra; so the trace in the quaternionic case must be defined (Finkelstein, Jauch, and Speiser, 1959) as $\text{Tr}\mathcal{O} = \text{Re} \sum_n \mathcal{O}_{nn}$, which does obey the cyclic properties of Eqs. (1.1a–c).

The main results of Chapters 1 through 6 generalize to the real and quaternionic cases, except for those that depend on the fact that the complex trace can have a non-zero imaginary part, or on the fact that in complex Hilbert space i acts as an anti-self-adjoint c-number. An example of the former is the derivation leading from Eq. (5.30a) to Eq. (5.33a), while examples of the latter are the discussions in Sections 4.5 and 5.5 that use the diagonalization of i_{eff} into $\pm i1_K$ sectors, and the part of the Brownian motion discussion of Section 6.1 that assumes a non-zero imaginary part $\mathcal{K}_1$ (which is needed for energy-driven reduction, but not for reduction by localization).

Appendix B: Algebraic proof of the Jacobi identity for the generalized Poisson bracket

We give here a basis-independent, algebraic proof of the Jacobi identity for the generalized Poisson bracket, following Adler, Bhanot, and Weckel (1994). For ease of exposition, we shall use a more compact notation than was employed in

Chapter 1. Derivatives with respect to q_r and p_r of a trace functional **A** will be denoted by $\mathbf{A}_r$ and $\mathbf{A}^r$ respectively. The operation Tr will be implied by a parenthesis ( ); this means that we can cyclically permute the factors within a parenthesis, if we include a factor ϵ_r every time a q_r or p_r is moved from the front of a parenthesis to the back, with $\epsilon_r = 1(-1)$ for bosonic (fermionic) degrees of freedom. Thus, in our shorthand notation, $(q_r\mathcal{O}) = \epsilon_r(\mathcal{O}q_r)$, and the generalized Poisson bracket is given by

$$\{\mathbf{A}, \mathbf{B}\} = \sum_r \epsilon_r \left(\mathbf{A}_r\mathbf{B}^r - \mathbf{B}_r\mathbf{A}^r\right). \tag{B.1}$$

It is useful to illustrate with an example how derivatives are computed. Consider the case where we have two kinds of matrix variables q_1, p_1 and q_2, p_2. Given the trace functional $\mathbf{A} = (q_1 p_1 q_2 q_1 p_2 q_1)$, its derivative with respect to q_1 is denoted by $\mathbf{A}_1$ and is given by

$$\mathbf{A}_1 = q_1 p_1 q_2 q_1 p_2 + \epsilon_1\epsilon_2 p_2 q_1 q_1 p_1 q_2 + \epsilon_1 p_1 q_2 q_1 p_2 q_1. \tag{B.2}$$

The three terms result from the three possible q_1 factors to differentiate, and the ϵ factors come from cyclically permuting the matrix factors to bring the particular q_1 which is to be differentiated to the right.

The first term on the left-hand side of the Jacobi identity of Eq. (1.13a), expanded out in this notation, is

$$\{\mathbf{A}, \{\mathbf{B}, \mathbf{C}\}\} = \sum_r \{\mathbf{A}, \epsilon_r \left(\mathbf{B}_r\mathbf{C}^r - \mathbf{C}_r\mathbf{B}^r\right)\}, \tag{B.3a}$$

which can be expanded further to

$$\{\mathbf{A}, \{\mathbf{B}, \mathbf{C}\}\} = \sum_{r,s} \epsilon_r\epsilon_s \left(\mathbf{A}_s \left(\mathbf{B}_r\mathbf{C}^r\right)^s - \mathbf{A}_s \left(\mathbf{C}_r\mathbf{B}^r\right)^s - \left(\mathbf{B}_r\mathbf{C}^r\right)_s \mathbf{A}^s + \left(\mathbf{C}_r\mathbf{B}^r\right)_s \mathbf{A}^s\right). \tag{B.3b}$$

Cyclic permutations of **A**, **B**, and **C** give the other two terms in Eq. (1.13a). Thus, the left-hand side of Eq. (1.13a) is

$$\sum_{r,s} \epsilon_r\epsilon_s[\left(\mathbf{A}_s \left(\mathbf{B}_r\mathbf{C}^r\right)^s - \mathbf{A}_s \left(\mathbf{C}_r\mathbf{B}^r\right)^s - \left(\mathbf{B}_r\mathbf{C}^r\right)_s \mathbf{A}^s + \left(\mathbf{C}_r\mathbf{B}^r\right)_s \mathbf{A}^s\right)$$
$$+ \left(\mathbf{B}_s \left(\mathbf{C}_r\mathbf{A}^r\right)^s - \mathbf{B}_s \left(\mathbf{A}_r\mathbf{C}^r\right)^s - \left(\mathbf{C}_r\mathbf{A}^r\right)_s \mathbf{B}^s + \left(\mathbf{A}_r\mathbf{C}^r\right)_s \mathbf{B}^s\right) \tag{B.4}$$
$$+ \left(\mathbf{C}_s \left(\mathbf{A}_r\mathbf{B}^r\right)^s - \mathbf{C}_s \left(\mathbf{B}_r\mathbf{A}^r\right)^s - \left(\mathbf{A}_r\mathbf{B}^r\right)_s \mathbf{C}^s + \left(\mathbf{B}_r\mathbf{A}^r\right)_s \mathbf{C}^s\right)].$$

Let us first consider how the terms in Eq. (B.4) cancel in the classical, *c*-number case. A similar cancellation mechanism will also apply in the more general matrix operator case. For *c*-numbers, the trace operation is trivial, derivatives of functionals commute, and one can apply the Leibniz product rule to expand the terms. For instance

$$\left(\mathbf{B}_r \mathbf{C}^r\right)^s = \mathbf{B}_r^{\,s} \mathbf{C}^r + \mathbf{B}_r \mathbf{C}^{rs}. \tag{B.5}$$

Note that $\mathbf{B}_r^{\,s}$ means that the q_r derivative is applied before the p_s derivative. $\mathbf{B}_{\,r}^s$ would mean that the same derivatives are applied in the opposite order. This distinction is meaningless for c-number fields, where derivatives commute, but it is crucial for noncommutative operators $\{q_r\}$ and $\{p_r\}$.

Equation (B.5) implies that each summand term in Eq. (B.4) will generate two terms. These terms cancel in pairs in the c-number case. For example, in the first term in Eq. (B.4), consider the derivative with respect to p_s applied to $\mathbf{B}_r$. This generates the term $+\mathbf{A}_s \mathbf{B}_r^{\,s} \mathbf{C}^r$. This cancels against the term $-\mathbf{A}_r \mathbf{B}_{\,s}^r \mathbf{C}^s$ obtained by applying the derivative with respect to p_s on $\mathbf{B}_r$ in the eleventh term (the dummy indices r and s need to be interchanged for the terms to be the same). The other half of the eleventh term will in turn be cancelled by a part of the eighth term, and so on. After twelve such double terms have been computed, we come back to the beginning and all terms have been cancelled.

The order in which these cancellations occur classically in the summand of Eq. (B.4) is as follows

$$\longleftrightarrow (\mathbf{A}_s\left(\mathbf{B}_r\mathbf{C}^r\right)^s) \longleftrightarrow ((\mathbf{A}_s\mathbf{B}^s)_r\,\mathbf{C}^r) \longleftrightarrow ((\mathbf{A}_r\mathbf{C}^r)_s\,\mathbf{B}^s) \longleftrightarrow (\mathbf{A}_r\left(\mathbf{C}_s\mathbf{B}^s\right)^r) \longleftrightarrow$$
$$(\mathbf{C}_s\left(\mathbf{A}_r\mathbf{B}^r\right)^s) \longleftrightarrow ((\mathbf{C}_s\mathbf{A}^s)_r\,\mathbf{B}^r) \longleftrightarrow ((\mathbf{C}_r\mathbf{B}^r)_s\,\mathbf{A}^s) \longleftrightarrow (\mathbf{C}_r\left(\mathbf{B}_s\mathbf{A}^s\right)^r) \longleftrightarrow$$
$$(\mathbf{B}_s\left(\mathbf{C}_r\mathbf{A}^r\right)^s) \longleftrightarrow ((\mathbf{B}_s\mathbf{C}^s)_r\,\mathbf{A}^r) \longleftrightarrow ((\mathbf{B}_r\mathbf{A}^r)_s\,\mathbf{C}^s) \longleftrightarrow (\mathbf{B}_r\left(\mathbf{A}_s\mathbf{C}^s\right)^r) \longleftrightarrow,$$
$$\tag{B.6}$$

where we have used the fact that r and s are dummy indices and have interchanged them in some of the terms, and where the lower right of Eq. (B.6) links back to the upper left. By Eq. (B.5), each entry in Eq. (B.6) generates two terms; one of these cancels against a term from the entry to the immediate left in the chain, and the other cancels against a term from the entry to the immediate right.

We will now proceed to show that in the general operator case, the cancellations occur in a similar way. However, the absence of both commutativity and the Leibniz product rule for operators makes the proof a little less trivial. For the rest of this discussion, we focus, as in Eq. (B.6), on the summands which appear, summed over r and s, in the Jacobi identity. Also, we will assume that $\mathbf{A}, \mathbf{B}, \mathbf{C}$ are monomials in $\{q_r\}$ and $\{p_r\}$. The proof for the general case of polynomial functionals follows from expanding out the generalized Poisson brackets in Eq. (1.13a) in terms of monomials.

When one computes the derivative of some monomial with respect to q_r (say), each particular occurrence of q_r generates one term in the result. Consider the expression

$$(\mathbf{B}_r \mathbf{C}^r)^s, \tag{B.7}$$

which appears in the first entry of Eq. (B.6). In this expression, there are three derivatives, and there is a sum over the set of choices of which occurrence of q_r, p_r, and p_s is differentiated in the appropriate factors. Each one of the set of choices will produce a particular monomial term in the result. If q_r appears $N(\mathbf{B}, q_r)$ times in the monomial $\mathbf{B}$, and p_r appears $N(\mathbf{C}, p_r)$ times in $\mathbf{C}$, and so on, then the number of terms produced by Eq. (B.7) is at most $N(\mathbf{B}, q_r)N(\mathbf{C}, p_r)[N(\mathbf{B}, p_s) + N(\mathbf{C}, p_s)]$.

We will show that in Eq. (B.4), each such monomial term in the result, for fixed r, s (i.e., for a fixed choice of q_r, p_r, q_s, p_s), will cancel with its counterpart in the order defined by Eq. (B.6). Consider the case where the p_s derivative is applied to $\mathbf{B}$ in the first entry and the q_r derivative is applied to $\mathbf{B}$ in the second entry of Eq. (B.6). For these to give non-vanishing contributions, $\mathbf{B}$ must contain at least one instance of both q_r and p_s. Therefore the most general form for $\mathbf{B}$ is

$$\mathbf{B} = (\alpha q_r \beta p_s), \tag{B.8}$$

where α and β are arbitrary monomials (and could possibly contain q_r and p_s). The displayed q_r and p_s are the particular instances of these coordinates in $\mathbf{B}$ upon which the derivatives will act.

We have

$$\begin{aligned}
(\mathbf{A}_s(\mathbf{B}_r \mathbf{C}^r)^s) &= (\mathbf{A}_s((\alpha q_r \beta p_s)_r \mathbf{C}^r)^s) \\
&= \epsilon_\alpha \epsilon_r (\mathbf{A}_s(\beta p_s \alpha \mathbf{C}^r)^s) \\
&= \epsilon_\alpha \epsilon_r \epsilon_\beta \epsilon_s (\mathbf{A}_s \alpha \mathbf{C}^r \beta), \tag{B.9}
\end{aligned}$$

and

$$\begin{aligned}
((\mathbf{A}_s \mathbf{B}^s)_r \mathbf{C}^r) &= ((\mathbf{A}_s(\alpha q_r \beta p_s)^s)_r \mathbf{C}^r) \\
&= ((\mathbf{A}_s \alpha q_r \beta)_r \mathbf{C}^r) \\
&= \epsilon_\beta (\beta \mathbf{A}_s \alpha \mathbf{C}^r) \\
&= (\mathbf{A}_s \alpha \mathbf{C}^r \beta). \tag{B.10}
\end{aligned}$$

If $\mathbf{B}$ is not identically zero (in which case the equality of Eqs. (B.9) and (B.10) is trivial), it must have an even number of fermion factors. Therefore, $\epsilon_\alpha \epsilon_r \epsilon_\beta \epsilon_s = 1$, and so the right-hand sides of Eqs. (B.9) and (B.10) are always the same. Finally, these same cancellations can be shown to occur for every summand term in Eq. (B.4) in the order indicated by Eq. (B.6), and apply both to the summands with $r \neq s$ and to those with $r = s$, including the parts of the summands with $r = s$

in which there are two derivatives with respect to the same variable q_r (or p_r). This proves that the Jacobi identity is true for arbitrary bosonic and fermionic matrix operator variables $\{q_r\}$ and $\{p_r\}$.

Appendix C: Symplectic structures in trace dynamics

We shall demonstrate here that there is a close correspondence (Adler and Wu, 1994) between the tangent vector field and symplectic structures of trace dynamics and of classical mechanics. (For expositions of the symplectic structure of classical mechanics, see Arnold, 1978 and Abraham and Marsden, 1980.) Let $X_\mathbf{A}$ be the tangent vector field associated with a trace functional $\mathbf{A}$, defined as a formal derivative operator by

$$X_\mathbf{A} \equiv \mathrm{Tr} \sum_r \left(\epsilon_r \frac{\delta \mathbf{A}}{\delta q_r} \frac{\delta}{\delta p_r} - \frac{\delta \mathbf{A}}{\delta p_r} \frac{\delta}{\delta q_r} \right), \tag{C.1}$$

which by definition acts on any trace functional $\mathbf{B}$ as

$$X_\mathbf{A} \mathbf{B} = \mathbf{B} X_\mathbf{A} + (X_\mathbf{A} \mathbf{B}), \tag{C.2}$$

with $(X_\mathbf{A} \mathbf{B})$ given by (cf. Eq. (1.11a))

$$(X_\mathbf{A} \mathbf{B}) = \mathrm{Tr} \sum_r \left(\epsilon_r \frac{\delta \mathbf{A}}{\delta q_r} \frac{\delta \mathbf{B}}{\delta p_r} - \frac{\delta \mathbf{A}}{\delta p_r} \frac{\delta \mathbf{B}}{\delta q_r} \right)$$

$$= \mathrm{Tr} \sum_r \epsilon_r \left(\frac{\delta \mathbf{A}}{\delta q_r} \frac{\delta \mathbf{B}}{\delta p_r} - \frac{\delta \mathbf{B}}{\delta q_r} \frac{\delta \mathbf{A}}{\delta p_r} \right) = \{\mathbf{A}, \mathbf{B}\}. \tag{C.3}$$

In terms of this operator, the time development of a general trace functional $\mathbf{B}[\{q_r\}, \{p_r\}]$ with no intrinsic time dependence, under the dynamics governed by $\mathbf{A}$ as trace Hamiltonian, can be written as (cf. Eq. (1.11b))

$$\frac{d\mathbf{B}}{dt} = -(X_\mathbf{A} \mathbf{B}). \tag{C.4}$$

Thus the tangent vector field $X_\mathbf{A}$ can be viewed as (minus) the directional derivative along the time evolution orbit (called the phase flow in the classical mechanics literature) of the phase space point $(\{q_r\}, \{p_r\})$, which is determined by the Hamiltonian equations of motion

$$\dot{p}_r = -\frac{\delta \mathbf{A}}{\delta q_r}, \quad \dot{q}_r = \epsilon_r \frac{\delta \mathbf{A}}{\delta p_r}, \tag{C.5}$$

with $\mathbf{A}$ acting as the trace Hamiltonian and with the dot denoting a time derivative. Following the terminology of classical mechanics, we call a tangent vector field of the form of Eq. (C.1) a Hamiltonian vector field. It is easily verified that the directional derivative X_A obeys the Leibniz product rule when applied to the generalized Poisson bracket

$$(X_A\{B, C\}) = \{(X_A B), C\} + \{B, (X_A C)\}, \tag{C.6}$$

because this equation is equivalent to the Jacobi identity of Eq. (1.13a).

Let us now study the algebraic structure of Hamiltonian vector fields, by computing the action of the commutator of two tangent vector fields X_A and X_B on a third trace functional C

$$\begin{aligned}([X_A, X_B]C) &= (X_A(X_B C)) - (X_B(X_A C)) \\ &= \{A, \{B, C\}\} - \{B, \{A, C\}\} \\ &= \{A, \{B, C\}\} + \{B, \{C, A\}\}. \end{aligned} \tag{C.7}$$

Using Eq. (C.3) with A replaced by $\{A, B\}$ and B replaced by C, we also get

$$(X_{\{A,B\}}C) = \{\{A, B\}, C\}\}, \tag{C.8}$$

and subtracting Eq. (C.8) from Eq. (C.7) gives finally

$$\begin{aligned}(([X_A, X_B] - X_{\{A,B\}})C) \\ = \{A, \{B, C\}\} + \{C, \{A, B\}\} + \{B, \{C, A\}\} = 0. \end{aligned} \tag{C.9}$$

Hence the validity of the Jacobi identity for the generalized Poisson bracket implies that the Hamiltonian vector fields X_A obey the commutator algebra

$$[X_A, X_B] = X_{\{A,B\}}, \tag{C.10}$$

and thus form a Lie algebra that is isomorphic to the Lie algebra of trace functionals under the generalized Poisson bracket. This gives a trace dynamics analog of a standard result in classical mechanics.

We next show that the symplectic geometry of classical mechanics extends to trace dynamics, and, as in the classical case, it is preserved by phase space flows produced by Hamiltonian time evolutions. Symplectic geometry is defined by an antisymmetric metric in the tangent or cotangent spaces of a phase space. (This contrasts with Riemannian geometry, which is defined by a symmetric metric in the tangent or cotangent spaces of a manifold.) To avoid differential forms, let us work in the cotangent space, which is spanned by covariant vectors, the components of which form the gradient of a function on phase space. The standard symplectic metric, or inner product, between two classical functions on phase space is provided by their classical Poisson bracket. The analogs of classical functions in trace dynamics are trace functionals, with differentials given by the phase space version of Eq. (1.3b)

$$\delta \mathbf{A} = \mathrm{Tr} \sum_r \left(\frac{\delta \mathbf{A}}{\delta q_r} \delta q_r + \frac{\delta \mathbf{A}}{\delta p_r} \delta p_r \right). \tag{C.11}$$

We can then use the generalized Poisson bracket to define a generalized symplectic structure Ω on the operator phase space, by defining the inner product between two cotangent vectors $\delta \mathbf{A}$ and $\delta \mathbf{B}$ by

$$\Omega(\delta \mathbf{A}, \delta \mathbf{B}) \equiv \{ \mathbf{A}, \mathbf{B} \}. \tag{C.12}$$

To see that this symplectic structure is preserved by the Hamiltonian dynamics given by Eq. (C.4), we observe that the time derivative of the inner product along the phase flow is

$$\frac{d}{dt} \Omega(\delta \mathbf{B}, \delta \mathbf{C}) = \frac{d}{dt} \{ \mathbf{B}, \mathbf{C} \} = \{ \{ \mathbf{B}, \mathbf{C} \}, \mathbf{A} \}, \tag{C.13a}$$

while that of the differential $\delta \mathbf{B}$ along the same flow is, by Eqs. (C.4) and (C.5)

$$\frac{d}{dt} \delta \mathbf{B} = \delta \dot{\mathbf{B}} = \delta \frac{d\mathbf{B}}{dt} = \delta \{ \mathbf{B}, \mathbf{A} \}. \tag{C.13b}$$

Therefore we have

$$\Omega(\delta \dot{\mathbf{B}}, \delta \mathbf{C}) + \Omega(\delta \mathbf{B}, \delta \dot{\mathbf{C}}) = \{ \dot{\mathbf{B}}, \mathbf{C} \} + \{ \mathbf{B}, \dot{\mathbf{C}} \}$$
$$= \{ \{ \mathbf{B}, \mathbf{A} \}, \mathbf{C} \} + \{ \mathbf{B}, \{ \mathbf{C}, \mathbf{A} \} \}, \tag{C.13c}$$

which comparing with Eq. (C.13a) and using the Jacobi identity of Eq. (1.13a) implies that

$$\frac{d}{dt} \Omega(\delta \mathbf{B}, \delta \mathbf{C}) = \Omega(\delta \dot{\mathbf{B}}, \delta \mathbf{C}) + \Omega(\delta \mathbf{B}, \delta \dot{\mathbf{C}}). \tag{C.14}$$

In other words, the symplectic structure is invariant under Hamiltonian phase flow. This statement can be viewed as a dual form of the Liouville theorem for trace dynamics.

Thus, trace dynamics, with non-commuting operator phase space variables, nonetheless has an underlying symplectic geometry which is preserved by the time evolution generated by any trace Hamiltonian, or equivalently, by the flow corresponding to any general canonical transformation as defined in Eq. (2.13a). Hence, in analogy with classical mechanics, the basic concepts and theorems of trace dynamics will be invariant under the group of transformations that preserve its generalized symplectic structure.

Appendix D: Gamma matrix identities for supersymmetric trace dynamics models

We give here the gamma matrix identities needed for carrying out the calculations involving supersymmetric trace dynamics models sketched in Sections 3.1, 3.2 and 3.3.

For the calculations of Sections 3.1 and 3.2, it is convenient to use Majorana representation γ matrices constructed explicitly as follows. Let $\sigma_{1,2,3}$ and $\tau_{1,2,3}$ be two independent sets of Pauli spin matrices; then we take

$$\gamma^0 = -\gamma_0 = -i\sigma_2\tau_1,$$
$$\gamma^1 = \gamma_1 = \sigma_3,$$
$$\gamma^2 = \gamma_2 = -\sigma_2\tau_2, \tag{D.1a}$$
$$\gamma^3 = \gamma_3 = -\sigma_1,$$
$$\gamma_5 = i\gamma^1\gamma^2\gamma^3\gamma^0 = -\sigma_2\tau_3,$$
$$\gamma^0\gamma_5 = \tau_2,$$

so that γ^0, γ_5, $\gamma^0\gamma_5$ are skew symmetric and $\gamma^{1,2,3}$ and $\gamma^0\gamma^{1,2,3}$ are symmetric, and

$$\gamma^0\gamma^{\mu T}\gamma^0 = \gamma^\mu. \tag{D.1b}$$

For this choice of γ matrices, the four matrices γ^μ are real.

To prove supersymmetry of the trace dynamics version of the Wess–Zumino model, one uses cyclic invariance of the trace together with the cyclic identity valid for Majorana representation γ matrices

$$\sum_{\text{cycle } a\to b\to d\to a} [\gamma^0_{ab}\gamma^0_{cd} + (\gamma^0\gamma_5)_{ab}(\gamma^0\gamma_5)_{cd}] = 0. \tag{D.2}$$

To prove supersymmetry of the trace dynamics version of the supersymmetric Yang–Mills model, one uses cyclic invariance of the trace together with another cyclic identity valid for Majorana representation γ matrices

$$\sum_{\text{cycle } a\to b\to d\to a} (\gamma^0\gamma^\mu)_{ab}(\gamma^0\gamma_\mu)_{cd} = 0. \tag{D.3}$$

To verify closure of the supersymmetry algebra under the generalized Poisson bracket, one can proceed in either of two ways. The first is to directly rearrange into the expected form, verifying along the way the various γ matrix identities that are needed; for example, in the case of the Wess–Zumino model, one needs the cyclic identity of Eq. (D.2) together with the additional identity (with ℓ, m, n

spatial indices, and $\epsilon_{\ell mn}$ the three index antisymmetric tensor with $\epsilon_{123} = 1$)

$$\gamma^\ell_{ab}\gamma^0_{cd} + \gamma^\ell_{db}\gamma^0_{ca} - (\gamma^\ell\gamma_5)_{ab}(\gamma^0\gamma_5)_{cd} - (\gamma^\ell\gamma_5)_{db}(\gamma^0\gamma_5)_{ca}$$
$$= \delta_{ad}(\gamma^0\gamma_\ell)_{bc} - (\gamma^0\gamma_\ell)_{ad}\delta_{bc} - i\sum_{m,n}\epsilon_{\ell mn}(\gamma_\ell\gamma_m\gamma_5)_{ad}(\gamma_\ell\gamma_n)_{cb}. \quad (D.4)$$

An alternative method for verifying the closure of the supersymmetry algebra is to first Fierz transform using the standard Fierz identity given in Eq. (A.80) of the book of West (1990), so as to isolate expressions of the form $\alpha^T\Gamma\beta$, and then to show that the coefficients of the various terms of this type, with different Dirac matrix structures Γ, have the form required by closure.

The identities of Eqs. (D.2–4) are representation covariant, in that they do not take the same form in representations in which the Dirac gamma matrices are complex rather than real. To see this, we note that the matrices in a general representation γ^μ_G are related to the Majorana representation matrices γ^μ given above by

$$\gamma^\mu_G = U^\dagger\gamma^\mu U = U^{T*}\gamma^\mu U, \quad (D.5)$$

with U a unitary matrix which in general is complex, as a result of which the row and column indices transform with different matrices. However, the identities of Eqs. (D.2–4) mix row and column indices; for example, in Eq. (D.2) there is one term in the cyclic sum in which a is a row index, and two terms in which a is a column index. (By way of contrast, the more familiar Fierz identities only interchange two row indices, and so do not mix row and column indices.) Hence we cannot get a representation invariant form of the identity by two applications of Eq. (D.5), since in the second and third terms of the cyclic sum, we will have a row index contracted with a U and a column index contracted with a U^*, which does not correspond with Eq. (D.5). However, we can easily get a representation covariant form of Eq. (D.2) by contracting all indices with a U^*, and wherever U^* contracts with a column index using the identity

$$U^* = UU^{*T}U^* = U\gamma^*, \gamma \equiv U^T U, \quad (D.6)$$

with γ a matrix that appears in the book of Adler (1995, pp. 341–342) (and which is introduced there because it plays a role in the transformation properties of the Dirac equation in quaternionic quantum mechanics). We can then apply Eq. (D.5) to all the gamma matrices, giving for Eq. (D.2), for example, the representation covariant form

$$\sum_{\text{cycle } a\to b\to d\to a} [(\gamma^0\gamma^*)_{ab}(\gamma^0\gamma^*)_{cd} + (\gamma^0\gamma_5\gamma^*)_{ab}(\gamma^0\gamma_5\gamma^*)_{cd}] = 0. \quad (D.7)$$

For a change of representation which preserves reality of the γ matrices, we have $U^* = U$, $\gamma = U^T U = U^{*T} U = 1$, and Eq. (D.7) is identical to Eq. (D.2), but for general changes of representation the identity is form covariant but not form invariant.

For the calculations in the "M Theory" model of Section 3.3, one uses a set γ_i of nine 16×16 matrices, that are related (Green, Schwartz, and Witten, 1987; Cremmer and Julia, 1979) to the standard 32×32 matrices Γ^μ as well as to the Dirac matrices of spin(8). A number of properties of the real, symmetric matrices γ_i play a role in the calculation. These matrices satisfy the anticommutator algebra

$$\{\gamma_i, \gamma_j\} = 2\delta_{ij}, \tag{D.8}$$

as well as the cyclic identity

$$\sum_{\text{cycle } p \to q \to n \to p} (\delta^{mn}\delta^{pq} - \gamma_i^{mn}\gamma_i^{pq}) = 0, \tag{D.9}$$

with i summed over and with the indices m, n, p, q spinorial indices ranging from 1 to 16. (The identity of Eq. (D.9) also has the same spinor index appearing both as a row and as a column index, and so is only form covariant under changes of gamma matrix representation, and is obtained by chiral projection with $\frac{1}{2}(1 - i\Gamma^9)$ from Eq. (4.A.6) of Green, Schwartz, and Witten, 1987.) Defining

$$\gamma_{ij} = \frac{1}{2}[\gamma_i, \gamma_j], \tag{D.10a}$$

so that

$$\gamma_i \gamma_j = \delta_{ij} + \gamma_{ij}, \tag{D.10b}$$

one readily derives from Eq. (D.9) an identity given by Banks, Seiberg, and Shenker (1997)

$$\gamma_{ij}^{mn}\gamma_i^{pq} + \gamma_{ij}^{pq}\gamma_i^{mn} + (m \leftrightarrow p) = 2(\gamma_j^{nq}\delta^{mp} - \gamma_j^{mp}\delta^{nq}), \tag{D.10c}$$

with i again summed over. By standard gamma matrix manipulations using Eq. (D.8), one also derives the fact that the matrix

$$A_{ijk} = \gamma_i \gamma_j \gamma_k - \delta_{ij}\gamma_k + \delta_{ik}\gamma_j - \delta_{jk}\gamma_i \tag{D.11}$$

is totally antisymmetric in the indices i, j, k (it is just the antisymmetrized product $\gamma_{[i}\gamma_j\gamma_{k]}$ with normalization factor $\frac{1}{6}$), as well as the identity

$$\frac{1}{2}\{\gamma_{\ell m}, \gamma_{ij}\} = \gamma_{[\ell}\gamma_m\gamma_i\gamma_{j]} + \delta_{\ell j}\delta_{im} - \delta_{mj}\delta_{i\ell}, \tag{D.12}$$

with the first term on the right the antisymmetrized product including normalization factor $\frac{1}{24}$.

Appendix E: Trace dynamics models with operator gauge invariance

In Section 3.2, we studied the trace dynamics version of the supersymmetric Yang–Mills model, which is an example of a general class of trace dynamics models with a local operator gauge invariance, and a corresponding operator constraint. In Section 4.4, we discussed methods for taking this constraint into account in forming the canonical ensemble and the partition function. Lest it appear that operator gauge invariance is intrinsically linked to supersymmetry, we give here further examples discussed by Adler (1994, 1995) of trace dynamics models, now non-supersymmetric, which admit an operator gauge invariance.

As our first example, we consider a matrix scalar field ϕ, which is not restricted to be self-adjoint (or anti-self-adjoint), and which is subjected to the general local gauging

$$\phi \to U\phi U'^{\dagger}, \qquad UU^{\dagger} = U^{\dagger}U = U'U'^{\dagger} = U'^{\dagger}U' = 1, \qquad \text{(E.1a)}$$

with U and U' independent unitary matrices. (Thus, the superscript $'$ (prime) in this Appendix does *not* have the significance of "non-commutative part" as in the discussion of Eqs. (2.16a–c) of Chapter 2.) Let us introduce independent anti-self-adjoint gauge potentials B_μ, B'_μ which transform as

$$B_\mu \to U B_\mu U^{\dagger} - (\partial_\mu U)U^{\dagger}, \qquad B'_\mu \to U' B'_\mu U'^{\dagger} - (\partial_\mu U')U'^{\dagger}, \qquad \text{(E.1b)}$$

and the covariant derivative and field strengths

$$\begin{aligned}
D_\mu \phi &= \partial_\mu \phi + B_\mu \phi - \phi B'_\mu, \\
F_{\mu\nu} &= \partial_\mu B_\nu - \partial_\nu B_\mu + [B_\mu, B_\nu], \\
F'_{\mu\nu} &= \partial_\mu B'_\nu - \partial_\nu B'_\mu + [B'_\mu, B'_\nu],
\end{aligned} \qquad \text{(E.1c)}$$

which correspondingly transform as

$$D_\mu \phi \to U D_\mu \phi U'^{\dagger}, \qquad F_{\mu\nu} \to U F_{\mu\nu} U^{\dagger}, \qquad F'_{\mu\nu} \to U' F'_{\mu\nu} U'^{\dagger}. \qquad \text{(E.1d)}$$

Then the trace Lagrangian density given by

$$\begin{aligned}
\mathcal{L} &= \mathcal{L}_\phi + \mathcal{L}_B + \mathcal{L}_{B'}, \\
\mathcal{L}_\phi &= \text{Tr}\left\{ \frac{1}{2}[-(D_\mu\phi)^{\dagger}D^\mu\phi - m^2\phi^{\dagger}\phi] - \frac{\lambda}{4}(\phi^{\dagger}\phi)^2 \right\}, \\
\mathcal{L}_B &= \text{Tr}\left(\frac{1}{4g^2} F_{\mu\nu}F^{\mu\nu} \right), \qquad \mathcal{L}_{B'} = \text{Tr}\left(\frac{1}{4(g')^2} F'_{\mu\nu}F'^{\mu\nu} \right),
\end{aligned} \qquad \text{(E.2a)}$$

is gauge invariant, as may be verified by substituting Eqs. (E.1a–d) and using cyclic invariance under the trace. The trace Lagrangian **L** and action **S** are formed from

$\mathcal{L}$ by the usual recipe

$$\mathbf{L} = \int d^3 x \mathcal{L}, \qquad \mathbf{S} = \int dt \, \mathbf{L}. \tag{E.2b}$$

When we form the Euler–Lagrange equations by varying $\mathbf{S}$, through $\delta F_{\mu\nu}$ and $\delta F'_{\mu\nu}$ we encounter new covariant derivatives $\hat{D}_\mu$ and $\hat{D}'_\mu$ defined by

$$\hat{D}_\mu \mathcal{O} = \partial_\mu \mathcal{O} + [B_\mu, \mathcal{O}], \qquad \hat{D}'_\mu \mathcal{O} = \partial_\mu \mathcal{O} + [B'_\mu, \mathcal{O}],$$
$$\delta F_{\mu\nu} = \hat{D}_\mu \delta B_\nu - \hat{D}_\nu \delta B_\mu, \qquad \delta F'_{\mu\nu} = \hat{D}'_\mu \delta B'_\nu - \hat{D}'_\nu \delta B'_\mu, \tag{E.3a}$$

and in integrating by parts we use the following "intertwining identities" that are easily derived from Eqs. (E.1c) and (E.3a)

$$\hat{D}_\mu(\rho\eta^\dagger) = (D_\mu\rho)\eta^\dagger + \rho(D_\mu\eta)^\dagger,$$
$$\hat{D}'_\mu(\rho^\dagger\eta) = (D_\mu\rho)^\dagger\eta + \rho^\dagger D_\mu\eta,$$
$$\partial_\mu \mathrm{Tr}\,(\rho\eta^\dagger) = \mathrm{Tr}\,[(D_\mu\rho)\eta^\dagger + \rho(D_\mu\eta)^\dagger],$$
$$\partial_\mu \mathrm{Tr}\,(\rho^\dagger\eta) = \mathrm{Tr}\,[(D_\mu\rho)^\dagger\eta + \rho^\dagger D_\mu\eta], \tag{E.3b}$$

which apply when ρ and η are both bosonic or both fermionic in type. We then get the operator equations of motion

$$D_\mu D^\mu \phi - (m^2 + \lambda\phi\phi^\dagger)\phi = 0,$$
$$\hat{D}^\mu F_{\mu\nu} = g^2 \mathcal{J}_\nu, \qquad \mathcal{J}_\nu = \frac{1}{2}[(D_\nu\phi)\phi^\dagger - \phi(D_\nu\phi)^\dagger], \tag{E.3c}$$
$$\hat{D}'^\mu F'_{\mu\nu} = g'^2 \mathcal{J}'_\nu, \qquad \mathcal{J}'_\nu = \frac{1}{2}[(D_\nu\phi)^\dagger\phi - \phi^\dagger D_\nu\phi],$$

in which the $\nu = 0$ components of the gauge field equations are constraints.

We turn next to the case of fermion fields, starting again with the operator gauging in which there is a fermion ψ transforming as

$$\psi \to U\psi U'^\dagger. \tag{E.4a}$$

The trace Lagrangian density analogous to Eqs. (E.2a) is

$$\mathcal{L} = \mathcal{L}_\psi + \mathcal{L}_B + \mathcal{L}_{B'}, \tag{E.4b}$$

with $\mathcal{L}_B$ and $\mathcal{L}_{B'}$ as in Eq. (E.2a), and with $\mathcal{L}_\psi$ given by

$$\mathcal{L}_\psi = \mathrm{Tr}(-\psi^\dagger\gamma^0\gamma^\mu D_\mu\psi + im\psi^\dagger\gamma^0\psi),$$
$$D_\mu\psi = \partial_\mu\psi + B_\mu\psi - \psi B'_\mu. \tag{E.4c}$$

This Lagrangian density is again gauge invariant, and varying the trace action $\mathbf{S}$ to get the corresponding Euler–Lagrange equations, we find the operator equations of motion

$$(-\gamma^\mu D_\mu + m)\psi = 0,$$

$$\hat{D}^\mu F_{\mu\nu} = g^2 \mathcal{J}_\nu, \qquad \mathcal{J}_\nu = \psi^T \gamma_\nu^T \gamma^{0T} \psi^{\dagger T}, \qquad\qquad \text{(E.4d)}$$

$$\hat{D}'^\mu F'_{\mu\nu} = (g')^2 \mathcal{J}'_\nu, \qquad \mathcal{J}'_\nu = \psi^\dagger \gamma^0 \gamma_\nu \psi,$$

with T indicating Dirac index (but *not* operator) transposition. Since ψ and $\psi^\dagger$ are non-commutative matrix operators, the current $\mathcal{J}_\nu$ is not equal to $-\mathcal{J}'_\nu$, as it would be if ψ, $\psi^\dagger$ were c-number Grassmann spinors. Again, the $\nu = 0$ components of the gauge field equations are constraints.

A further discussion of the bosonic and fermionic models briefly described here, including their Hamiltonian form and their discrete symmetries, can be found in Adler (1994, 1995).

Appendix F: Properties of Wightman functions needed for reconstruction of local quantum field theory

We review here, following Streater and Wightman (1968), the properties of Wightman functions that are needed for the reconstruction from them of local quantum field theory. For simplicity, we consider only the case of a single self-adjoint scalar field $\phi(x)$. Letting $|\text{vac}\rangle$ denote the vacuum state, which is assumed unique, the Wightman functions are defined by

$$\mathcal{W}(x_1, x_2, \ldots, x_n) = \langle \text{vac} | \phi(x_1)\phi(x_2) \ldots \phi(x_n) | \text{vac} \rangle, \qquad \text{(F.1)}$$

for all n ranging from 0 to ∞. Starting from the axioms of local quantum field theory, a number of properties of these functions can be derived. Conversely, given Wightman functions satisfying the following properties, one can reconstruct a local quantum field theory:

(i) Smoothness properties
The functions $\mathcal{W}(\{x\})$ must be tempered distributions.

(ii) Covariance
The functions $\mathcal{W}(\{x\})$ must satisfy the requirements of Poincaré invariance. Translation invariance requires that

$$\mathcal{W}(x_1, x_2, \ldots, x_n) = W(x_1 - x_2, x_2 - x_3, \ldots, x_{n-1} - x_n)$$

$$\equiv W(\xi_1, \xi_2, \ldots, \xi_{n-1}) \equiv W(\{\xi\}). \qquad \text{(F.2)}$$

Lorentz invariance for a scalar field ϕ requires that

$$W(\{\xi\}) = W(\{\Lambda\xi\}),\tag{F.3}$$

with Λ_μ^ν a proper orthochronous Lorentz transformation. When fields with spin appear in the Wightman functions, Eq. (F.3) must be modified to include the appropriate Wigner rotations acting on the spin indices.

(iii) Spectral condition

Let $\tilde{W}(p_1, p_2, \ldots, p_n)$ and $\tilde{W}(q_1, q_2, \ldots, q_{n-1})$ be the Fourier transforms of the Wightman functions defined by

$$\tilde{W}(p_1, p_2, \ldots, p_n) = \int dx_1 \ldots dx_n \exp\left(-i \sum_{j=1}^{n} p_j \cdot x_j\right) W(x_1, x_2, \ldots, x_n),$$

$$\tilde{W}(q_1, q_2, \ldots, q_{n-1}) = \int d\xi_1 \ldots d\xi_{n-1} \exp\left(-i \sum_{j=1}^{n-1} q_j \cdot \xi_j\right) W(\xi_1, \xi_2, \ldots, \xi_{n-1}).$$

$$\tag{F.4}$$

These must be related by

$$\tilde{W}(p_1, p_2, \ldots, p_n) = (2\pi)^4 \delta\left(\sum_{j=1}^{n} p_j\right)$$

$$\times \tilde{W}(p_1, p_1 + p_2, \ldots, p_1 + p_2 + \ldots + p_{n-1}), \quad \text{(F.5a)}$$

and we must have

$$\tilde{W}(q_1, q_2, \ldots, q_{n-1}) = 0,\tag{F.5b}$$

for any q_j not in the forward light cone defined by $q^0 \geq |\vec{q}\,|$. (We have followed here the notation of Streater and Wightman (1968); in the notation that we have used in Chapters 1–7, dx would be written as d^4x and $\delta(p)$ as $\delta^4(p) = \delta(p^0)\delta^3(\vec{p})$.)

(iv) Local commutativity

The Wightman functions must obey

$$W(x_1, \ldots, x_j, x_{j+1}, \ldots, x_n) = W(x_1, \ldots, x_{j+1}, x_j, \ldots, x_n),$$

$$j = 1, 2, \ldots, n-1,\tag{F.6}$$

whenever x_j and x_{j+1} are spacelike separated.

(v) Hermiticity and positivity conditions

The Wightman functions must obey the hermiticity condition

$$\mathcal{W}(x_1, \ldots, x_n) = \mathcal{W}(x_n, \ldots, x_1)^*, \tag{F.7a}$$

where $*$ denotes the complex conjugate. They must also obey the positivity condition

$$\sum_{j,k} \int \ldots \int dx_1 \ldots dx_j dy_1 \ldots dy_k f_j(x_1, \ldots, x_j)^*$$

$$\times \mathcal{W}(x_j, \ldots, x_1, y_1, \ldots, y_k) f_k(y_1, \ldots, y_k) \geq 0, \tag{F.7b}$$

for all finite sequences $f_0, f_1(x_1), f_2(x_1, x_2), \ldots$ of test functions.

(vi) Cluster property

The Wightman functions must cluster, in the sense that

$$\lim_{S \to \infty} [\mathcal{W}(x_1, \ldots, x_j, x_{j+1} + Sa, \ldots, x_n + Sa)$$

$$- \mathcal{W}(x_1, \ldots, x_j) \mathcal{W}(x_{j+1}, \ldots, x_n)] = 0, \tag{F.8}$$

when the unit four-vector direction a of increasing separation S is spacelike,

The proof of the reconstruction theorem, assuming Wightman functions obeying the conditions enumerated above, is given in Streater and Wightman (1968), and a recent discussion is given in Strocchi (1993).

Appendix G: BRST invariance transformation for global unitary fixing

In this Appendix we formulate a BRST invariance transformation associated with the global unitary fixing given in Section 4.6, again following the treatment given in Adler and Horwitz (2003). Our starting point is an alternative ghost representation for the De Witt–Faddeev–Popov determinant Δ, obtained by using Eq. (4.59c) for the factors associated with the matrix B to give

$$\Delta \propto \int d\omega d\tilde{\omega} \exp \left(\text{Tr} \tilde{\omega}[\omega, A] + \sum_{j=2}^{K} \tilde{\omega}_{jj} (\text{Re} B_{1j}) i (\omega_{jj} - \omega_{11}) \right). \tag{G.1}$$

Yet another equivalent form for Δ is obtained by noting that

$$[B, \omega]_{1j} = B_{1j}(\omega_{jj} - \omega_{11}) + S, \tag{G.2a}$$

with the remainder S denoting terms that only involve matrix elements ω_{ij} with $i \neq j$. The remainder S makes a vanishing contribution to the Grassmann integrals when Eq. (G.2a) is substituted for $B_{1j} i (\omega_{jj} - \omega_{11})$ in Eq. (G.1), since one factor of $(\omega_{jj} - \omega_{11})$ for each $j = 2, \ldots, K$ is needed to give a non-vanishing integral, and each such term in the exponent is already accompanied by a factor $\tilde{\omega}_{jj}$, so that terms with additional such factors vanish inside the Grassmann integrals. (We are just using here the fact that with ζ, $\tilde{\zeta}$ Grassmann variables, $\int d\zeta d\tilde{\zeta} \exp(\tilde{\zeta} W \zeta + U \zeta) = \int d\zeta d\tilde{\zeta} \exp(\tilde{\zeta} W \zeta + \tilde{\zeta} U) = W$, with no dependence on U.) Since the diagonal matrix elements of ω are pure imaginary, Eq. (G.2a) implies that

$$(\mathrm{Re}\, B_{1j}) i (\omega_{jj} - \omega_{11}) = -\mathrm{Im}[B, \omega]_{1j}, \tag{G.2b}$$

which when substituted into Eq. (G.1) gives the alternative formula

$$\Delta \propto \int d\omega d\tilde{\omega} \exp\left(\mathrm{Tr}\tilde{\omega}[\omega, A] - \sum_{j=2}^{K} \tilde{\omega}_{jj} \mathrm{Im}[B, \omega]_{1j}\right). \tag{G.3}$$

We will also need the Fourier representations of Eqs. (4.62a–c) for the delta function and step function constraints. An alternative form of the step function constraint of Eq. (4.62c) is obtained by including in the exponent an additional term $-\sum_{j=2}^{K} \kappa_j \mathrm{Re}[B, \omega]_{1j}$, with κ_j auxiliary Grassmann parameters that are *not* integrated over. This term is linear in ω but does not involve $\tilde{\omega}$, and so by the same argument as in the preceding paragraph, it makes a vanishing contribution when the step function constraint is substituted into the overall formula for $\mathcal{J}$ and the Grassmann integrals over ω and $\tilde{\omega}$ are carried out. Using the alternative forms of Δ and of the step function constraint to rewrite Eq. (4.63), we get a new representation for the unitary-fixed integral

$$\mathcal{J} = C \int dM dh dH dk d\omega d\tilde{\omega}$$

$$\times \exp\left(i \mathrm{Tr} h A + \mathrm{Tr}\tilde{\omega}[\omega, A] + \sum_{j=2}^{K}(i H_j \mathrm{Im} B_{1j} - \tilde{\omega}_{jj} \mathrm{Im}[B, \omega]_{1j}\right.$$

$$\left. + i k_j \mathrm{Re} B_{1j} - \kappa_j \mathrm{Re}[B, \omega]_{1j})\right) \mathcal{G}[\{M\}], \tag{G.4}$$

with C an overall constant factor. This representation of $\mathcal{J}$ will be used to establish a BRST invariance.

We now show that Eq. (G.4) is manifestly invariant under the nilpotent BRST transformation

$$\delta A = [A, \omega]\theta, \quad \delta B = [B, \omega]\theta, \quad \delta M_d = [M_d, \omega]\theta, \ d = 3, \ldots, D,$$

$$\delta \omega = \omega^2 \theta, \quad \delta \tilde{\omega}_{ij} = -i h_{ij}\theta, \ i \neq j, \quad \delta \tilde{\omega}_{jj} = -i H_j\theta, \ j = 2, \ldots, K, \tag{G.5}$$

$$\delta h = 0, \quad \delta H_j = 0, \quad \delta k_j = 0, \quad \delta \kappa_j = -i k_j\theta,$$

with θ a c-number Grassmann parameter. (The part of this transformation involving ω is patterned after the BRST transformation for the local operator gauge invariant case studied by Adler (1998) and used in Section 4.4.) We first remark that since Eq. (G.5) has the form of an infinitesimal unitary transformation acting on the matrix variables M_d with generator $\omega\theta$ that is anti-self-adjoint (since $(\omega\theta)^\dagger = -\theta^\dagger\omega^\dagger = \theta\omega = -\omega\theta$), the global unitary invariant function $\mathcal{G}[\{M\}]$ and the matrix integration measure dM are both invariant. We consider next the terms in the exponent in Eq. (G.4). From Eq. (G.5) we have

$$\delta[A, \omega] = [\delta A, \omega] + [A, \delta\omega] = [[A, \omega]\theta, \omega] + [A, \omega^2\theta]$$
$$= -(\omega[A, \omega] + [A, \omega]\omega)\theta + [A, \omega^2]\theta = -[A, \omega^2]\theta + [A, \omega^2]\theta = 0.$$

$$\text{(G.6a)}$$

Hence for the terms in the exponent of Eq. (G.4) involving A, we get

$$\delta(i\,\mathrm{Tr}h A + \mathrm{Tr}\tilde{\omega}[\omega, A]) = i\,\mathrm{Tr}h\delta A + \mathrm{Tr}(\delta\tilde{\omega})[\omega, A]$$
$$= i\,\mathrm{Tr}h[A, \omega]\theta + \mathrm{Tr}(-ih\theta)[\omega, A] = 0. \qquad\text{(G.6b)}$$

For the terms in the exponent of Eq. (G.4) involving B but not involving the parameters k_j inside the summation over j we have

$$\delta(i H_j \mathrm{Im} B_{1j} - \tilde{\omega}_{jj}\mathrm{Im}[B, \omega]_{1j}) = i H_j \mathrm{Im}\delta B_{1j} - (\delta\tilde{\omega}_{jj})\mathrm{Im}[B, \omega]_{1j}$$
$$= i H_j \mathrm{Im}[B, \omega]_{1j}\theta + i H_j\theta\mathrm{Im}[B, \omega]_{1j} = 0, \qquad\text{(G.6c)}$$

since $\delta[B, \omega] = 0$ by the same argument as in Eq. (G.6a). Finally, the terms in the exponent involving the step function parameters k_j and κ_j are invariant by a similar argument. So the entire exponent of the representation in Eq. (G.4) is manifestly BRST invariant.

Continuing the BRST analysis, since $\mathrm{Tr}\sigma\tau = -\mathrm{Tr}\tau\sigma$ for any two Grassmann odd grade matrices τ and σ, we have $\mathrm{Tr}\omega^2 = -\mathrm{Tr}\omega^2 = 0$, and so the condition that ω should be traceless is preserved by Eq. (G.5). (On the other hand, ω_{11}^2 is non-zero even when ω_{11} is zero, which is why we must use a traceless condition, rather than a condition $\omega_{11} = 0$, for ω.) Also, since

$$(\omega^2)^\dagger = -(\omega^\dagger)^2 = \omega^2, \qquad\text{(G.6d)}$$

the property that ω is anti-self-adjoint is preserved by Eq. (G.5). The property that $\tilde{\omega}$ is anti-self-adjoint is preserved by Eq. (G.5) because the matrix h is self-adjoint. The integration measures dh and dH are trivially invariant, while the measure $d\tilde{\omega}$

is invariant because $\delta\tilde{\omega}$ has no dependence on $\tilde{\omega}$. Since

$$\delta(d\omega_{ij}) = d(\delta\omega)_{ij} = d(\omega^2\theta)_{ij} = \left(\omega d\omega + (d\omega)\omega\right)_{ij}\theta, \tag{G.7a}$$

we have

$$\delta(d\omega_{ij}) = (\omega_{ii}d\omega_{ij} + d\omega_{ij}\omega_{jj})\theta + \dots = d\omega_{ij}(\omega_{jj} - \omega_{ii})\theta + \dots \tag{G.7b}$$

with ... denoting terms that contain only matrix elements $d\omega_{i'j'}$ with $(i', j') \neq (i, j)$. Hence there is no Jacobian contribution from the diagonal terms in the measure $d\omega$, while the Jacobian arising from transformation of the off-diagonal terms in $d\omega$ differs from unity by a term proportional to

$$\sum_{i \neq j}(\omega_{jj} - \omega_{ii})\theta = 0, \tag{G.7c}$$

and so the measure $d\omega$ is also invariant. Finally, nilpotence of the BRST transformation follows from Eq. (G.6a), and its analogs with A replaced by B or by a general M_d, together with

$$\delta\omega^2 = \{\delta\omega, \omega\} = \{\omega^2\theta, \omega\} = \omega^2\{\theta, \omega\} = 0. \tag{G.7d}$$

This completes the demonstration of BRST invariance under Eq. (G.5) of the representation of the unitary-fixed integral given in Eq. (G.4).

References

Numbers in square brackets following each reference indicate the pages where the reference is cited.

Abraham, R. and J. E. Marsden (1980). *Foundations of Mechanics*, 2nd edn (Reading, MA: Benjamin/Cummings), p. 194. [198]

Adler, S. L. (1979). Algebraic chromodynamics. *Phys. Lett. B* **86**, 203–205. [19]

(1994). Generalized quantum dynamics. *Nucl. Phys. B* **415**, 195–242. [19,24,29,204,206]

(1995). *Quaternionic Quantum Mechanics and Quantum Fields* (New York: Oxford University Press), Secs. 13.5–13.7 and App. A. [19,24,202,204,206]

(1997a). Poincaré supersymmetry representations over trace class non-commutative graded operator algebras. *Nucl. Phys. B* **499**, 569–582. [20,64]

(1997b). The matrix model for M theory as an examplar of trace (or generalized quantum) dynamics. *Phys. Lett. B* **407**, 229–233. [20,64,72]

(1997c). A strategy for a vanishing cosmological constant in the presence of scale invariance breaking. *Gen. Rel. and Grav.* **29**, 1357–1362. [61,191]

(1998). Gauge fixing in the partition function for generalized quantum dynamics. *J. Math. Phys.* **39**, 1723–1729. [94,208]

(2000). Remarks on a proposed Super–Kamiokande test for quantum gravity induced decoherence effects. *Phys. Rev. D* **62**, 117901. [179]

(2002). Environmental influence on the measurement process in stochastic reduction models. *J. Phys. A: Math. Gen.* **35**, 841–858. [169,182,183]

(2003a). Weisskopf–Wigner decay theory for the energy-driven stochastic Schrödinger equation. *Phys. Rev. D* **67**, 025007. [169,176,177,179]

(2003b). Why decoherence has not solved the measurement problem: a response to P. W. Anderson. *Stud. Hist. Philos. Mod. Phys.* **34**, 135–142. [4]

Adler, S. L., G. V. Bhanot, and J. D. Weckel (1994). Proof of Jacobi identity in generalized quantum dynamics. *J. Math. Phys.* **35**, 531–535. [19,29,194]

Adler, S. L., D. C. Brody, T. A. Brun, and L. P. Hughston (2001). Martingale models for quantum state reduction. *J. Phys. A: Math. Gen* **34**, 8795–8820. [169,173]

Adler, S. L. and T. A. Brun (2001). Generalized stochastic Schrödinger equations for state vector collapse. *J. Phys. A: Math. Gen.* **34**, 1–13. [167,169]

Adler, S. L. and L. P. Horwitz (1996). Microcanonical ensemble and algebra of conserved generators for generalized quantum dynamics. *J. Math. Phys.* **37**, 5429–5443. [19,31,45,46,88]

(2000). Structure and properties of Hughston's stochastic extension of the Schrödinger equation. *J. Math. Phys.* **41**, 2485–2499. [169,170,182]

(2003). Global unitary fixing and matrix-valued correlations in matrix models. *Phys. Lett. B.* **570**, 73–81. [20,106,208]

Adler, S. L. and A. Kempf (1998). Corrections to the emergent canonical commutation relations arising in the statistical mechanics of matrix models. *J. Math. Phys.* **39**, 5083–5097. [19,20,43,102,144,145]

Adler, S. L. and A. C. Millard (1996). Generalized quantum dynamics as pre-quantum mechanics. *Nucl. Phys. B* **473**, 199–244. [19,20,33,43,45,76,117,130]

Adler, S. L. and Y.-S. Wu (1994). Algebraic and geometric aspects of generalized quantum dynamics. *Phys. Rev. D* **49**, 6705–6708. [30,198]

Arndt, M., O. Nairz, J. Vos-Andreae, C. Keller, G. van der Zouw, and A. Zeilinger (1999). Wave-particle duality of C_{60} molecules. *Nature* **401**, 680–682. [188]

Arnold, V. I. (1978). *Mathematical Methods of Classical Mechanics* (New York; Springer-Verlag), p. 211. [198]

Ax, J. and S. Kochen (1999). Extension of quantum mechanics to individual systems. arXiv: quant-ph/9905077. Revised version in preparation. [8]

Banks, T., W. Fischler, S. H. Shenker, and L. Susskind (1997). M theory as a matrix model: a conjecture. *Phys. Rev. D* **55**, 5112–5128. [70]

Banks, T., N. Seiberg, and S. Shenker (1997). Branes from matrices. *Nucl. Phys. B* **490**, 91–106. [70,203]

Bassi, A. and G. C. Ghirardi (2002). Dynamical reduction models with general Gaussian noise. *Phys. Rev. A* **65**, 042114. [169]

(2003). Dynamical reduction models. *Physics Reports* **379**, 257–426. [17,169,188]

Bedford, D. and D. Wang (1975). Toward an objective interpretation of quantum mechanics. *Nuovo Cim.* **26**, 313–325. [169]

(1977). A criterion for spontaneous state reduction. *Nuovo Cim.* **37**, 55–62. [169]

Bell, J. S. (1964). On the Einstein–Podolsky–Rosen paradox. *Physics* **1**, 195–200. [6,7,17,152,153]

(1987). *Speakable and Unspeakable in Quantum Mechanics* (Cambridge: Cambridge University Press). [6,152]

Berezin, F. A. (1966). *The Method of Second Quantization* (London: Academic Press). [193]

Bergshoeff, E., E. Sezgin, and P. K. Townsend (1987). Supermembranes and eleven-dimensional supergravity. *Phys. Lett. B* **189**, 75–78. [70]

(1988). Properties of the eleven-dimensional supermembrane theory. *Ann. Phys.* **185**, 330–368. [70]

Bialynicki-Birula, I. and Mycielski, J. (1976). Nonlinear wave mechanics. *Ann. Phys.* **100**, 62–93. [18]

Bjorken, J. D. and S. D. Drell (1965). *Relativistic Quantum Fields* (New York: McGraw-Hill), Sec. 11.4. [55]

Bohm, D. (1952). A suggested interpretation of the quantum theory in terms of "hidden" variables, I and II. *Phys. Rev.* **85**, 166–193. [7]

Bollinger, J. J., D. J. Heinzen, W. M. Itano, S. L. Gilbert, and D. J. Wineland (1991). Atomic physics tests of nonlinear quantum mechanics, in *Atomic Physics 12: Proceedings of the 12th International Conference on Atomic Physics*, ed. J. C. Zorn and R. R. Lewis (New York: American Institute of Physics Press). [18]

Born, M. and P. Jordan (1925). Zur quantenmechanik. *Zeit. f. Phys.* **34**, 858–888 (see pp. 861–863). [18,24]

Bousso, R. (2002). The holographic principle. *Rev. Mod. Phys.* **74**, 825–874. [142,192]

Brézin, E. and S. R. Wadia (ed.) (1993). *The Large N Expansion in Quantum Field Theory and Statistical Mechanics* (Singapore: World Scientific). [15,75,106]

Brody, D. C. and L. P. Hughston (2002). Efficient simulation of quantum state reduction. *J. Math. Phys.* **43**, 5254–5261. [180]

Bub, J. (1997). *Interpreting the Quantum World* (Cambridge: Cambridge University Press). [8]

Callan, C., S. Coleman, and R. Jackiw (1970). A new improved energy-momentum tensor. *Ann. Phys.* **59**, 42–73. [58]

Cheng, T.-P. and L.-F. Li (1988). *Gauge Theory of Elementary Particle Physics* (Oxford: Clarendon Press). [193]

Claudson, M. and Halpern, M. B. (1985). Supersymmetric ground state wave functions. *Nucl. Phys. B* **250**, 689–715. [70]

Clauser, J. F., M. A. Horne, A. Shimony, and R. A. Holt (1969). Proposed experiment to test local hidden-variable theories. *Phys. Rev. Lett.* **23**, 880–884. [153]

Coleman, S. (1989). *Aspects of Symmetry* (Cambridge: Cambridge University Press), pp. 394–395. [111]

Collett, B., P. Pearle, F. Avignone, and S. Nussinov (1995). Constraint on collapse models by limit on spontaneous X-ray emission in Ge. *Found. Phys.* **25**, 1399–1412. [189]

Cook, R. J. (1988). What are quantum jumps? *Phys. Scr. T* **21** 49–51. [179]

Cremmer, E. and B. Julia (1979). The $SO(8)$ supergravity. *Nucl. Phys. B* **159**, 141–212. [203]

de Wit, B. (1997). Supersymmetric quantum mechanics, supermembranes and Dirichlet particles. *Nucl. Phys. B Proc. Suppl.* **56B**, 76–87. *Theory of Elementary Particles*, ed. D. Lüst, H.-J. Otto, and G. Weigt (Amsterdam: North-Holland). [70]

de Wit, B., J. Hoppe, and H. Nicolai (1988). On the quantum mechanics of supermembranes. *Nucl. Phys. B* **305**, 545–581. [70]

De Witt, B. S. and N. Graham (1973). *The Many-Worlds Interpretation of Quantum Mechanics* (Princeton, NJ: Princeton University Press). [8]

Diósi, L. (1988a). Quantum stochastic processes as models for state vector reduction. *J. Phys. A: Math. Gen.* **21**, 2885–2898. [17,169]

(1988b). Continuous quantum measurement and Itô formalism. *Phys. Lett. A* **129**, 419–423. [17,169]

(1989). Models for universal reduction of macroscopic quantum fluctuations. *Phys. Rev. A* **40**, 1165–1174. [17,169]

Dürr, D., S. Goldstein, R. Tumulka, and N. Zanghi (2003). Quantum Hamiltonians and stochastic jumps. arXiv: quant-ph/0303056. [8]

Dürr, D., S. Goldstein, and N. Zanghi (1992). Quantum equilibrium and the origin of absolute uncertainty. *J. Stat. Phys.* **67**, 843–907. [8]

Einstein, A., B. Podolsky, and N. Rosen (1935). Can quantum-mechanical description of physical reality be considered complete? *Phys. Rev.* **47**, 777–780. [10]

Everett, H., III (1957). "Relative State" formulation of quantum mechanics. *Rev. Mod. Phys.* **29**, 454–462. [8]

Facchi, P. and S. Pascazio (2003). Unstable systems and quantum Zeno phenomena in quantum field theory, in *Fundamental Aspects of Quantum Mechanics*, ed. L. Accardi and S. Tasaki (Singapore: World Scientific). [177]

Feynman, R. P., F. L. Vernon, Jr., and R. W. Hellwarth (1957). Geometrical representation of the Schrödinger equation for solving maser problems. *J. Appl. Phys.* **28**, 49–52. [177]

Finkelstein, D., J. M. Jauch, and D. Speiser (1959). Notes on quaternion quantum mechanics. CERN Report 59-7 (unpublished). Reprinted in *Logico-Algebraic Approach to Quantum Mechanics II*, ed. C. Hooker (Dordrecht: Reidel, 1979). [194]

Fivel, D. I. (1997). An indication from the magnitude of CP violation that gravitation is a possible cause of wave-function reduction. arXiv: quant-ph/9710042. [169]

Flume, R. (1985). On quantum mechanics with extended supersymmetry and nonabelian gauge constraints. *Ann. Phys.* **164**, 189–220. [70]

Friedman, J. R., V. Patel, W. Chen, S. K. Tolpygo, and J. E. Lukens (2000). Quantum superposition of distinct macroscopic states. *Nature* **406**, 43–45. [178]

Gardiner, C. W. (1990). *Handbook of Stochastic Methods* (Berlin: Springer-Verlag), Chap. 4. [164]

Ghirardi, G. C., P. Pearle, and A. Rimini (1990). Markov processes in Hilbert space and continuous spontaneous localization of systems of identical particles. *Phys. Rev. A* **42**, 78–89. [17,168,169,170,186,188]

Ghirardi, G. C., A. Rimini, and T. Weber (1986). Unified dynamics for microscopic and macroscopic systems. *Phys. Rev. D* **34**, 470-491. [17,169]

Gisin, N. (1984). Quantum measurements and stochastic processes. *Phys. Rev. Lett.* **52**, 1657–1660. [17,169,173]

    (1989). Stochastic quantum dynamics and relativity. *Helv. Phys. Acta* **62**, 363–371. [17,18,168,169]

    (1990). Weinberg's non-linear quantum mechanics and supraluminal communications. *Phys. Lett. A* **143**, 1–2. [18,168,169]

Gisin, N. and M. Rigo (1995). Relevant and irrelevant nonlinear Schrödinger equations. *J. Phys. A: Math. Gen.* **28**, 7375–7390. [18,168]

Giulini, D. (2003). That strange procedure called quantization, in *Aspects of Quantum Gravity – From Theory to Experimental Search*, Bad Honnef 2002 School, ed. D. Giulini, C. Kiefer, and C. Lämmerzahl (Berlin: Springer-Verlag). [10,30]

Giulini, D., E. Joos, C. Kiefer, J. Kupsch, J. O. Stamatescu, and H. D. Zeh (1996). *Decoherence and the Appearance of a Classical World in Quantum Theory* (Berlin: Springer-Verlag), Sec. 3.3.1. [177]

Gorini, V., A. Kossakowski, and E. C. G. Sudarshan (1976). Completely positive dynamical semigroups of N-level systems. *J. Math. Phys.* **17**, 821–825. [168]

Green, M. B., J. H. Schwarz, and E. Witten (1987). *Superstring Theory*, Vol. I (Cambridge: Cambridge University Press), pp. 220, 246, and 288. [71,203]

Greenberg, W. R., A. Klein, I. Zlatev, and C.-T. Li (1996, revised 2001). From Heisenberg matrix mechanics to EBK quantization: theory and first applications. arXiv: chem-ph/9603006. [18]

Griffiths, R. B. (2002). *Consistent Quantum Theory* (Cambridge: Cambridge University Press). [7]

Gürsey, F. and C.-H. Tze (1996). *On the Role of Division, Jordan and Related Algebras in Particle Physics* (Singapore: World Scientific), pp. 38–39. [138]

Harris, R. A. and L. Stodolsky (1981). On the time dependence of optical activity. *J. Chem. Phys.* **74**, 2145–2155. [4]

Hartle, J. B. (1992). Spacetime quantum mechanics and the quantum mechanics of spacetime, in *Gravitation et Quantifications*, Les Houches summer school 1992, ed. B. Julia and J. Zinn-Justin (Amsterdam: Elsevier, 1995), pp. 285–480. [7]

Hughston, L. P. (1996). Geometry of stochastic state vector reduction. *Proc. Roy. Soc. Lond. A* **452**, 953–979. [169,170,171,174,181]

Itano, W. M., D. J. Heinzen, J. J. Bollinger, and D. J. Wineland (1990). Quantum Zeno effect. *Phys. Rev. A* **41**, 2295–2300. [179]

Joos, E. (1999). Elements of environmental decoherence, in *Decoherence: Theoretical, Experimental, and Conceptual Problems*, ed. P. Blanchard, D. Giulini, E Joos, C. Kiefer, and I.-O. Stamatescu (New York: Springer-Verlag), pp. 1–17. [4]

Joos, E. and H. D. Zeh (1985). The emergence of classical properties through interaction with the environment. *Z. Phys. B* **59**, 223–243. [4]

Kaku, M. (1993). *Quantum Field Theory* (New York: Oxford University Press), pp. 407–410. [118]

Kerman, A. and A. Klein (1963). Generalized Hartree-Fock approximation for the calculation of collective states of a finite many-particle system. *Phys. Rev.* **132**, 1326–1342, App. B. [18]

Klein, A., C. T. Li, and M. Vassanji (1980). Variational principles for particles and fields in Heisenberg matrix mechanics. *J. Math. Phys.* **21**, 2521–2527. [18]

Kloeden, P. E. and E. Platen (1997). *Numerical Solutions of Stochastic Differential Equations* (Berlin: Springer-Verlag). [180]

Kochen, S. and E. P. Specker (1967). The problem of hidden variables in quantum mechanics. *J. Math. Mech.* **17**, 59–87. [6,7,17,152]

Lindblad, G. (1976). On the generators of quantum dynamical semigroups. *Commun. Math. Phys.* **48**, 119–130. [168]

Lisi, E., A. Marrone, and D. Montanino (2000). Probing possible decoherence effects in atmospheric neutrino experiments. *Phys. Rev. Lett.* **85**, 1166–1169. [179]

Mehta, M. L. (1967). *Random Matrices* (New York: Academic Press), Chapt. 3. [107]

Milburn, G. J. (1991). Intrinsic decoherence in quantum mechanics. *Phys. Rev. A* **44**, 5401–5406. [169]

Millard, A. C. (1997). Non-Commutative methods in quantum mechanics, Princeton University Ph.D. thesis, Sec. 3.3. [19,43]

Misra, B. and E. C. G. Sudarshan (1977). The Zeno's paradox in quantum theory. *J. Math. Phys.* **18**, 756–763. [177]

Mohling, F. (1982). *Statistical Mechanics: Methods and Applications* (New York: Halsted Press/John Wiley), pp. 270–272. [118]

Moroz, I., R. Penrose, and P. Tod (1998). Spherically-symmetric solutions of the Schrödinger-Newton equations. *Class. Quantum Grav.* **15**, 2733–2742. [185]

Nairz, O., M. Arndt, and A. Zeilinger (2000). Experimental challenges in fullerene interferometry. *J. Mod. Optics* **47**, 2811–2821. [188]

Nairz, O., B. Brezger, M. Arndt, and A. Zeilinger (2001). Diffraction of complex molecules by structures made of light. *Phys. Rev. Lett.* **87**, 160401. [188]

Nelson, E. (1969). Derivation of the Schrödinger equation from Newtonian mechanics. *Phys. Rev.* **150**, 1079–1085. [18]

(1985). *Quantum Fluctuations* (Princeton, NJ: Princeton University Press). [18]

Ng, Y. J. (2003). Selected topics in Planck-scale physics. arXiv: gr-qc/0305019, Sec. V. [185]

Omnès, R. (1992). Consistent interpretations of quantum mechanics. *Rev. Mod. Phys.* **64**, 339–382. [7]

(1994). *The Interpretation of Quantum Mechanics* (Princeton, NJ: Princeton University Press). [7]

(1999). *Understanding Quantum Mechanics* (Princeton, NJ: Princeton University Press). [7]

Pearle, P. (1976). Reduction of the state vector by a nonlinear Schrödinger equation. *Phys. Rev. D* **13**, 857–868. [17,169]

(1979). Toward explaining why events occur. *Int. Journ. Theor. Phys.* **18**, 489–518. [17,169]

(1982). Might God toss coins? *Found. Phys.* **12**, 249–263. [169,172]

(1984). Experimental tests of dynamical state-vector reduction. *Phys. Rev. D* **29**, 235–240. [17,169]

(1989). Combining stochastic dynamical state-vector reduction with spontaneous localization. *Phys. Rev. A* **39**, 2277–2289. [17,168,169,172]

(1990). Toward a relativistic theory of statevector reduction, in *Sixty-Two Years of Uncertainty: Historical, Philosophical, and Physical Inquiries into the Foundations of Quantum Mechanics*, ed. A. I. Miller (New York: Plenum Press). [169]

(1993). Wavefunction collapse models with nonwhite noise, in *Perspectives on Quantum Reality*, ed. R. Clifton (Dordrecht: Kluwer), pp. 93–109. [169]

(1995). Ways to describe dynamical state-vector reduction. *Phys. Rev. A* **48**, 913–923. [169]

(1999a). Relativistic collapse model with tachyonic features. *Phys. Rev. A* **59**, 80–101. [169]

(1999b). Collapse models, in *Open Systems and Measurements in Relativistic Quantum Field Theory*, Lecture Notes in Physics 526, ed. H.-P. Breuer and F. Pettrucione (Berlin: Springer-Verlag). [17,169]

(2003). Problems and aspects of energy-driven wavefunction collapse models. arXiv: quant-ph/0310086. *Phys. Rev. A* **69** (in press). [182,184,185]

Pearle, P. and E. Squires (1994). Bound state excitation, nucleon decay experiments, and models of wave function collapse. *Phys. Rev. Lett* **73**, 1–5. [189]

Pearle, P., J. Ring, J. I. Collar, and F. T. Avignone III (1999). The CSL collapse model and spontaneous radiation: an update. *Found. Phys.* **29**, 465–480. [189]

Penrose, R. (1996). On gravity's role in quantum state reduction. *Gen. Rel. Grav.* **28**, 581–600. [185]

Percival, I. (1994). Primary state diffusion. *Proc. Roy. Soc. London A* **447**, 189–209. [17,169]

(1995). Quantum spacetime fluctuations and primary state diffusion. *Proc. Roy. Soc. A* **451**, 503–513. [169]

(1998). *Quantum State Diffusion* (Cambridge: Cambridge University Press). [169,180]

Peres, A. (1993). *Quantum Theory: Concepts and Methods* (Dordrecht: Kluwer), pp. 196–201. [152]

Polchinski, J. (1991). Weinberg's nonlinear quantum mechanics and the Einstein–Podolsky–Rosen paradox. *Phys. Rev. Lett.* **66**, 397–400. [18,168]

Rabi, I. I. (1937). Space quantization in a gyrating magnetic field. *Phys. Rev.* **51**, 652–654. [177]

Rabi, I. I., N. F. Ramsey, and J. Schwinger (1954). Use of rotating coordinates in magnetic resonance problems. *Rev. Mod. Phys.* **26**, 167–171. [177]

Rittenberg, V. and S. Yankielowicz (1985). Supersymmetric gauge theories in quantum mechanics. *Ann. Phys.* **162**, 273–302. [70]

Sculley, M. O., B.-G. Englert, and J. Schwinger (1989). Spin coherence and Humpty-Dumpty. III. The effects of observation. *Phys. Rev. A* **40**, 1775–1784. [3]

Shack, R. and T. A. Brun (1997). A C++ library using quantum trajectories to solve quantum master equations. *Comp. Phys. Commun.* **102**, 210–228. [180]

Shack, R., T. A. Brun, and I. C. Percival (1995). Quantum state diffusion, localization and computation. *J. Phys. A: Math. Gen.* **28**, 5401–5413. [180]

Smolin, L. (1983). Derivation of quantum mechanics from a deterministic non-local hidden variables theory, I. The two dimensional theory. Institute for Advanced Study preprint. [18,75]

(1985). Stochastic mechanics, hidden variables and gravity, in *Quantum Concepts in Space and Time*, ed. C. J. Isham and R. Penrose (New York: Oxford University Press, 1986). [18,75]

(2002). Matrix models as non-local hidden variables theories. arXiv: hep-th/0201031. Published in *Fukuoka 2001, String theory*, pp. 244–261. [18,75]

Sommerfeld, A. (reprinted 1956). *Thermodynamics and Statistical Mechanics* (New York: Academic Press), Secs. 28, 29, 36, and 40. [85]

Starodubtsev, A. (2002). A note on quantization of matrix models. arXiv: hep-th/0206097. [19]

Streater, R. F. and A. S. Wightman (1968). *PCT, Spin and Statistics, and all That* (New York: Benjamin). [136,206–208]

Strocchi, F. (1993). *Selected Topics on the General Properties of Quantum Field Theory*, Lecture Notes in Physics Vol. 51 (Singapore: World Scientific). [208]

Taylor, W. (2001). M(atrix) theory: matrix quantum mechanics as a fundamental theory. *Rev. Mod. Phys.* **73**, 419–462. [70,190]

ter Haar, D. (1995). *Elements of Statistical Mechanics*, third edn (Oxford and Boston, MA: Butterworth Heinemann), Sec. 5.13. [85]

't Hooft, G. (1988). Equivalence relations between deterministic and quantum mechanical systems. *J. Stat. Phys.* **53**, 323–344. [18]

(1997). Quantummechanical behavior in a deterministic model. *Found. Phys. Lett.* **10**, 105–111. [18]

(1999a). Quantum gravity as a dissipative deterministic system. *Class. Quant. Grav.* **16**, 3263–3279. [18]

(1999b). Determinism and dissipation in quantum gravity, in *Basics and Highlights in Fundamental Physics*, Erice 1999 Summer School, ed. A. Zichichi (Singapore: World Scientific, 2001), pp. 397–413. [18]

(2001a). How does God play dice? (Pre-)determinism at the Planck scale, in *Quantum [Un]speakables, from Bell to Quantum Information*, ed. R. A. Bertlmann and A. Zeilinger (Berlin: Springer-Verlag, 2002), pp. 307–316. [18]

(2001b). Quantum mechanics and determinism, in *Proceedings of the Eighth Int. Conf. on Particles, Strings and Cosmology*, ed. P. Frampton and J. Ng (Princeton, NJ: Rinton Press), pp. 275–285. [18]

(2002). Determinism beneath quantum mechanics. arXiv: quant-ph/0212095. To appear in *Proceedings of Quo Vadis Quantum Mechanics*, Temple University, PA, September 25, 2002. [18]

(2003). Determinism in free bosons. *Int. Journ. Theoret. Phys.* **42**, 355–361. [18]

Townsend, P. K. (1996). D-branes from M-branes. *Phys. Lett. B* **373**, 68–75. [70]

Vafa, C. and E. Witten (1983). Restrictions on symmetry breaking in vector-like gauge theories. *Nucl. Phys. B* **234**, 173–188. [86]

(1984). Parity conservation in quantum chromodynamics. *Phys. Rev. Lett.* **53**, 535–536. [86]

van der Wal, C. H., A. C. J. ter Haar, F. K. Wilhelm, R. N. Schouten, C. J. P. M. Harmans, T.P. Orlando, S. Lloyd, and J. E. Mooij (2000). Quantum superposition of macroscopic persistent-current states. *Science* **290**, 773–777. [178]

Vion, D., A. Aassime, A. Cottet, P. Joyez, H. Pothier, C. Urbina, D. Esteve, and M. H. Devoret (2002). Manipulating the quantum state of an electrical circuit. *Science* **296**, 886–889. [178]

von Neumann, J. (1932). *Mathematische grundlagen der quantenmechanik* (Berlin: Springer-Verlag). Engl. transl.: R. T. Beyer (1971). *Mathematical Foundations of*

Quantum Mechanics (Princeton, NJ: Princeton University Press), Chapter VI. [3,181]

Wadia, S. R. (1981). Integration over a large random matrix by Faddeev–Popov method, Chicago preprint EFI 81/37, archived at the Stanford Linear Accelerator library. [111]

Weber, T. (1990). Quantum mechanics with spontaneous localization revisited. *Nuovo Cimento B* **106**, 1111–1124. [186]

Weinberg, S. (1989a). Particle states as realizations (linear or nonlinear) of spacetime symmetries. *Nucl. Phys. B Proc. Suppl.* **6**, 67–75. *Spacetime Symmetries*, ed. Y. S. Kim and W. W. Zachary (Amsterdam: North-Holland). [18]

    (1989b). Precision tests of quantum mechanics. *Phys. Rev. Lett.* **62**, 485–488. [18]

    (1989c). Testing quantum mechanics. *Ann. Phys.* **194**, 336–386. [18]

    (1995). *The Quantum Theory of Fields*, Vol. I (Cambridge: Cambridge University Press), Sec. 7.4. [55]

    (1996). *The Quantum Theory of Fields*, Vol. II (Cambridge: Cambridge University Press), Sec. 15.5. [94,97,109]

Weinstein, S. (2001). Naive quantum gravity, in *Physics Meets Philosophy at the Planck Scale*, ed. C. Callender and N. Huggett (Cambridge: Cambridge University Press). [74]

West, P. (1990). *Introduction to Supersymmetry and Supergravity*, extended second edn (Singapore: World Scientific). [64,72,73,198,202]

Wheeler, J. A. and W. H. Zurek (ed.) (1983). *Quantum Theory and Measurement* (Princeton, NJ: Princeton University Press). [3]

Yoneya, T. and H. Itoyama (1982). Internal collective motions and the large-N limit in field theories. *Nucl. Phys. B* **200**, 439–456. [111]

Yu, Y., S. Han, X. Chu, S.-I. Chu, and Z. Wang (2002). Coherent temporal oscillations of macroscopic quantum states in a Josephson junction. *Science* **296**, 889–892. [178]

Zurek, W. H. (1991). Decoherence and the transition from quantum to classical. *Phys. Today* **44** (10), 36–44. [4]

Zurek, W. H. (2003). Decoherence, einselection, and the quantum origins of the classical. *Rev. Mod. Phys.* **75**, 715–775. [4]

Index